Fluence Monitoring in UV Disinfection Systems: Development of a Fluence Meter

Awwa
Research
Foundation

About the Awwa Research Foundation

The Awwa Research Foundation (AwwaRF) is a member-supported, international, nonprofit organization that sponsors research to enable water utilities, public health agencies, and other professionals to provide safe and affordable drinking water to consumers.

The Foundation's mission is to advance the science of water to improve the quality of life. To achieve this mission, the Foundation sponsors studies on all aspects of drinking water, including supply and resources, treatment, monitoring and analysis, distribution, management, and health effects. Funding for research is provided primarily by subscription payments from approximately 1,000 utilities, consulting firms, and manufacturers in North America and abroad. Additional funding comes from collaborative partnerships with other national and international organizations, allowing for resources to be leveraged, expertise to be shared, and broad-based knowledge to be developed and disseminated. Government funding serves as a third source of research dollars.

From its headquarters in Denver, Colorado, the Foundation's staff directs and supports the efforts of more than 800 volunteers who serve on the board of trustees and various committees. These volunteers represent many facets of the water industry, and contribute their expertise to select and monitor research studies that benefit the entire drinking water community.

The results of research are disseminated through a number of channels, including reports, the Web site, conferences, and periodicals.

For subscribers, the Foundation serves as a cooperative program in which water suppliers unite to pool their resources. By applying Foundation research findings, these water suppliers can save substantial costs and stay on the leading edge of drinking water science and technology. Since its inception, AwwaRF has supplied the water community with more than $300 million in applied research.

More information about the Foundation and how to become a subscriber is available on the Web at **www.awwarf.org**.

Fluence Monitoring in UV Disinfection Systems: Development of a Fluence Meter

Prepared by:
Guus F. IJpelaar and **Danny Harmsen**
Kiwa N.V. Water Research
Groningenhaven 7, P.O. Box 1072
3430 BB Nieuwegein, The Netherlands

Charles M. Sharpless and **Karl G. Linden**
Duke University, Department of Civil Engineering
P.O. Box 90287, Durham, NC 27708-0287

and

Joop C. Kruithof
PWN Water Supply Company North Holland
Rijksweg 501, P.O. Box 2113
1990 AC Velserbroek, The Netherlands

Jointly sponsored by:
Awwa Research Foundation
6666 West Quincy Avenue, Denver, CO 80235-3098

and

Kiwa Water Research
Groningenhaven 7, Postbus 1072, 3430 BB Nieuwegein, The Netherlands

Published by:

Awwa Research Foundation

American Water Works Association

IWA Publishing

DISCLAIMER

This study was jointly funded by the Awwa Research Foundation (AwwaRF) and Kiwa Water Research (Kiwa). AwwaRF and Kiwa assume no responsibility for the content of the research study reported in this publication or for the opinions or statements of fact expressed in the report. The mention of trade names for commercial products does not represent or imply the approval or endorsement of AwwaRF or Kiwa. This report is presented solely for informational purposes.

CONTENTS

TABLES

FIGURES

FOREWORD

The Awwa Research Foundation (AwwaRF) is a nonprofit corporation that is dedicated to the implementation of a research effort to help utilities respond to regulatory requirements and traditional high-priority concerns of the industry. The research agenda is developed through a process of consultation with subscribers and drinking water professionals. Under the umbrella of a Strategic Research Plan, the Research Advisory Council prioritizes the suggested projects based upon current and future needs, applicability, and past work; the recommendations are forwarded to the Board of Trustees for final selection. The foundation also sponsors research projects through an unsolicited proposal process; the Collaborative Research, Research Applications, and Tailored Collaboration programs; and various joint research efforts with organizations such as the U.S. Environmental Protection Agency, the U.S. Bureau of Reclamation, and the Association of California Water Agencies.

This publication is a result of one of these sponsored studies, and it is hoped that its findings will be applied in communities throughout the world. The following report serves not only as a means of communicating the results of the water industry's centralized research program but also as a tool to enlist the further support of the nonmember utilities and individuals.

Projects are managed closely from their inception to the final report by the foundation's staff and large cadre of volunteers who willingly contribute their time and expertise. The foundation serves a planning and management function and awards contracts to other institutions such as water utilities, universities, and engineering firms. The funding for this research effort comes primarily from the Subscription Program, through which water utilities subscribe to the research program and make an annual payment proportionate to the volume of water they deliver and consultants and manufacturers subscribe based on their annual billings. The program offers a cost-effective and fair method for funding research in the public interest.

A broad spectrum of water supply issues is addressed by the foundation's research agenda: resources, treatment and operations, distribution and storage, water quality and analysis, toxicology, economics, and management. The ultimate purpose of the coordinated effort is to assist water suppliers to provide the highest possible quality of water economically and reliably. The true benefits are realized when the results are implemented at the utility level. The foundation's trustees are pleased to offer this publication as a contribution toward that end.

Walter J. Bishop, P.E.
Chair, Board of Trustees
Awwa Research Foundation

Robert C. Renner, P.E.
Executive Director
Awwa Research Foundation

ACKNOWLEDGMENTS

Kiwa Water Research acknowledges the contribution of Gertjan Medema and Erwin Beerendonk in this project. The team at Duke University thanks Ki Don Cho for his assistance in parts of the study.

The authors also acknowledge the input and guidance provided by Awwa Research Foundation's senior project manager, Kenan Ozekin, and the members of the Project Advisory Committee; Robert A. Hulsey from Black & Veatch; and Harold Wright and Robert S. Cushing from Carollo Engineers.

EXECUTIVE SUMMARY

Worldwide, the application of ultraviolet (UV) light is becoming an accepted way of disinfecting water. As the sensitivity of microorganisms towards UV light may vary strongly, the required UV dose varies as well. The decimal elimination capacity describing the inactivation of the microorganisms is highly sensible towards the delivered UV dose (or UV fluence). For this reason, UV systems must be designed in such a way that all water flowing through the UV reactor should receive the minimum required UV dose. Therefore, short circuits in water flow through the reactor should be avoided as much as possible. To realize properly disinfected water, the UV dose must be evident at any time. UV sensors are available that enable the UV dose to be monitored. However, these sensors only detect 254 nm of UV light, which is unsatisfactory for UV systems equipped with medium-pressure lamps emitting UV light in the 200–300-nm range. Moreover, these sensors are inaccurate, break down easily, and must be recalibrated frequently. Generally, each UV lamp has its own UV sensor. Therefore, the readings of each individual sensor must be combined with computational fluid dynamics to enable calculation of the UV dose. To overcome these disadvantages, the research described in this report focused on the photolysis of a chemical substance giving a linear relation between the formation level of the photo product and the applied dose.

RESEARCH OBJECTIVES

One objective of this research was to develop an online fluence meter based on a UV-sensitive chemical for which the formation level of the photo product has a clear relation with the applied UV dose. Another objective was to find naturally occurring substances or chemicals to be dosed that would be sensitive towards UVC light (200–285 nm) and would give accurately measurable photo products in the 0–1,000 J/m^2 dose range.

APPROACH

The chemicals producing a clear and mathematically useful relationship between the concentration of the photo product and the applied UV dose may be either naturally occurring compounds or toxicologically acceptable chemicals that have to be dosed. The chemicals selected were investigated by performing collimated beam tests. Although several parameters will affect the UV dose in UV disinfection systems, only the variation in water quality was taken into account in this research.

Based on an initial study, nitrate as a naturally occurring compound and the ferric ion/ferrous ion (Fe^{3+}/Fe^{2+}) and hydrogen peroxide (H_2O_2)/benzoate systems as chemicals to be dosed were selected for further research.

CONCLUSIONS

Nitrate

Applying UV doses up to 1,000 J/m^2, the formation of nitrite as the photo product of nitrate has a linear relation with the applied UV dose at 3–14 mg/L nitrate levels. This offers the opportunity to determine the UV dose by analyzing the nitrite content of the water leaving the UV reactor.

However, the nitrite formation varies with the water quality. This implies that the contribution of the water quality to the level of nitrite must be known and quantified. In addition to the nitrite concentration in the reactor effluent, the initial nitrate content of the water and the pH should be monitored to correct the nitrite data for changes in water quality. Changes in nitrate levels can be accurately monitored measuring the UV absorbance over the 200–240-nm range, provided that the changes in dissolved organic carbon (DOC) content are only minor. Nitrite formation by photolysis of nitrate takes place in the 200–240-nm range. A very small effect on nitrite formation was ascribed to changes in DOC, which parameter absorbs UV light predominantly in the 200–240/250-nm range. However, because this effect was hardly measurable and the DOC content of most natural waters changes only slightly in time, effects of changes in DOC on nitrite formation can be neglected in most cases. Besides the initial nitrate concentration, the pH of the water must be known to determine the nitrite formation at a given water quality. Finally, the delivered UV dose can be calculated once the relation between the nitrite formation and the applied dose are determined. A monitoring system that is based on the relationship between the nitrite formation and the UV dose, including corrections for the water quality, is proposed. Both nitrite analysis and pH measurement are inexpensive and can be performed online. UV absorbance scans (200–300 nm) can be recorded online as well but require some investment costs.

Water disinfection (e.g., inactivation of microorganisms by DNA damage) is a process taking place in the 240–280-nm range. This range differs significantly from the wavelength range required for nitrate–nitrite conversion. As a result of the "short" wavelengths, nitrate absorbs UV light at relatively low distances from the UV lamp. This means that nitrite formation takes place in certain distinct volumes of water near each lamp. However, the "long" germicidal 240–280-nm wavelengths travel deeper through the water. Based on this discrepancy, the UV dose calculated using the level of nitrite formation may differ from the germicidal UV dose, which, based on current knowledge, may lead to deviations in the calculated UV dose.

The Fe^{3+}/Fe^{2+} and the H_2O_2/Benzoate Systems

The Fe^{3+}/Fe^{2+} and the H_2O_2/benzoate systems are both capable of quantifying germicidal fluences. The required sensitivity in natural water was found for the Fe^{3+}/Fe^{2+} system but not for the H_2O_2/benzoate system. To some extent, a nonlinear relationship exists between the fluence and the photo product at the higher fluences for both systems.

The colorimetric analysis of Fe^{2+} as the photo product is relatively inexpensive, and the expertise is easy to find. Sensors for online detection of Fe^{2+} have been described in the literature. In contrast, the maximum working pH is 6, which makes the iron system less practical.

The H_2O_2/benzoate system has no pH restriction. However, because the hydroxybenzoate photo products are analyzed by high-performance liquid chromatography, the analysis method requires a high level of expertise and is therefore costly. Also, this advanced analysis makes online detection rather impossible.

The use of H_2O_2 and benzoate may be of concern with respect to drinking water standards. This may not be an issue for iron as it appeared that doses as low as 1 mg/L Fe^{3+} can be used. Although the necessary concentrations of benzoate and the formation level of the hydroxy benzoate photo product are expected to be low enough to eliminate toxicity concerns, this item needs to be addressed.

RECOMMENDATIONS

Nitrate

In contrast to DNA, the required wavelengths for the production of hydroxyl radicals with the UV/H_2O_2 oxidation process take place in more or less the same wavelength range as the nitrate–nitrite conversion. Therefore, the discrepancy between DNA and nitrate with respect to the required wavelengths establishing germicidal fluences may not play a role compared with the nitrate fluence meter for oxidation purposes with UV light combined with H_2O_2. The UV/H_2O_2 advanced oxidation process can be characterized as a source of hydroxyl radicals ($OH^{\bullet}$) for the degradation of organic micropollutants. The UV/H_2O_2 oxidation process can be applied as a universal barrier against pathogens and undesired micropollutants. It is expected that there will be growing interest for this process, which implies that fluence monitoring for oxidation processes will become increasingly important.

CHAPTER 1
INTRODUCTION

BENEFITS FOR WATER UTILITIES

To guarantee the required disinfection credits during ultraviolet (UV) disinfection, it is essential to monitor the UV fluence in practical UV systems. Sensors are commonly used for fluence measurement, and they are based on detection of 254-nm UV radiation. The sensors are calibrated in a certain operational window, which is a function of the lamp intensity (ageing, fouling), water flow, and water transmission. Detection of the 254-nm UV radiation takes place at a certain point in the reactor, which data are used to calculate the fluence. However, this provides no information about the fluence delivered to the water experienced by microorganisms in the UV reactor. Therefore, the readings of each individual sensor (most UV units have one sensor per lamp) must be combined with computational fluid dynamics to calculate the reduction equivalent fluence (REF). The REF links the disinfection kinetics of microorganisms to the disinfection credits of the UV system. The currently applied method using sensors determines the REF of the reactor based on the batch-wise inactivation of calibrated microorganisms (biodosimetry). Subsequently, the data from the biodosimetry tests are used to calibrate the sensors. However, until now, the calculated REFs were not very reliable.

To enable monitoring of the average germicidal fluence in UV systems equipped with medium-pressure mercury (Hg) lamps, determination of REF values based on biodosimetry is currently under development. However, this approach is more complicated than that for low-pressure lamps. In 2002 it was recognized that a fluence meter for UV plants in operation is not available. During the international UV workshop at Kiwa Water Research in the Netherlands, a fluence meter was prioritized as the most important research item (Kiwa Water Research 2002). Therefore, research has been performed focusing on the development of a monitor for online fluence measurement. The fluence meter studied in this research is defined as a UVC-sensitive dissolved compound that is easily converted to its photo product under the operating conditions required for UV disinfection of natural water. Moreover, by enabling the determination of the fluence, a clear relationship between the applied fluence and the concentration of the photo product should exist.

In this research, both compounds of natural origin and dosed compounds were examined. Although the chance of finding and developing a fluence meter based on a compound to be dosed is relatively high, application of a naturally occurring compound has distinct advantages:

- There are no costs with respect to addition of a compound. Furthermore, there are no equipment and protocols necessary for addition of the compound.
- The presence of the compound is generally accepted in water sources that involve drinking water production (toxicity issue).
- Its presence guarantees a homogeneous distribution, which makes mixing equipment unnecessary.

Compared with the currently applied sensors and the biodosimetry method, the advantage of an online fluence meter is the continuous monitoring of the fluence and control of the diode-based sensors, which makes the process of monitoring more reliable.

CRITERIA FOR SELECTION OF COMPOUNDS

Compounds were selected for further research according to the following criteria:

- Compounds should be sensitive for UVC radiation, resulting in detectable product formation or changes in concentrations or structure of the original compound.
- The target photo product should be formed only by direct UV photolysis of the original compound and preferably is not a conversion product or formed through parallel occurring processes, such as, for example, biological processes.
- The target photo product should be the result of direct UV photolysis without the need for another compound(s) in the water acting as an intermediate compound in the photolysis process.
- The photolytic conversion should follow zero-order kinetics.
- Other oxidants (e.g., O_2) should not significantly interfere with the photolysis process.
- The analytical method(s) should enable measuring the target photo product with a detection limit in the parts-per-billion range and an analytical error <5%.
- Sensors for online monitoring of the promising compounds should be or become commercially available. These sensors should support the application of a fluence meter based on the analytical method, the detection limit, and the measurement and sample frequency.
- If the required sensors are commercially unavailable, in-house development of such sensors for online monitoring should still be feasible for the compound to be selected.

Enabling the selection of optional compounds to be developed as a fluence meter, the following parameters were used to characterize the compounds:

- Absorption range: With respect to the development of a monitor displaying the REF, the compound should preferably absorb UV light of the same wavelengths as that of DNA.
- Absorption coefficient ε: Although a high quantum yield is not necessarily linked to a high ε, a certain value for ε is required to obtain a high quantum yield.
- Quantum yield Φ: The value for Φ determines the yield of the photo product.
- Presence in natural waters: As described previously, there are certain advantages of a naturally occurring compound versus a compound to be dosed.
- Sensor for online detection: This is required for online monitoring.
- Product analysis (detection limit and accuracy): The lower the concentrations and the more accurate the target photo product can be analyzed, the more precise will be the results of the fluence meter.

APPROACH AND STRATEGY

This report describes the setup of the research for developing a fluence meter for online monitoring of the fluence in practical UV systems. The approach is based on the extent to which a distinct photo product is formed during UV irradiation of the parent compound in UV disinfection units. Therefore, compounds are selected by producing a clear and mathematically useful relationship between the concentration of the photo product in the effluent of the UV reactor and the applied fluence (0–1,000 J/m^2). The approach for the selection of compounds is based on naturally occurring compounds (research by Kiwa Water Research) and compounds to be dosed (research by Duke University).

The fundamental aspects required for the development of a chemical fluence meter were investigated by performing collimated beam tests. Effects of water quality on the performance of the fluence meter have been taken into account. Although process parameters such as water flow and lamp power affect the average fluence, these parameters were not included in this study.

In the development of a fluence meter, the final system may range from an alarm system that warns the end users of unacceptable low fluences to a full monitoring system that displays the REF, indicating the system disinfection efficacy. In the development of a full monitor, the fluence meter would be additional to or in replacement of biodosimetry. In addition, the fluence meter may enable end users to check the performance of the sensors installed by the manufacturer.

The research described in this report focuses on the development of a fluence meter applicable for UV disinfection systems. However, since application of a practical fluence meter is also evident for other UV technologies, the results are also considered for processes as the upcoming UV/hydrogen peroxide (H_2O_2) process.

SCOPE OF THIS REPORT

Chapter 2 presents the outcome of the literature research selecting compounds optional for further research into the development of a fluence meter. The background information of the selected compounds is given in chapter 3. Chapter 4 describes the materials and methods used throughout this research. The results and discussion are described in chapter 5. Based on the results, a monitoring system is proposed for practical application in chapter 6. Finally, the conclusions and some recommendations of this research are given in chapter 7.

CHAPTER 2
SELECTION OF COMPOUNDS

An overview of organic and inorganic substances tentatively selected for further research based on the parameters described was discussed in chapter 1. Test results performed with some of these compounds are discussed in the literature (IJpelaar and Kruithof 2003). Based on literature and the results of some preliminary tests, the formation of nitrite by UV photolysis of nitrate is a promising process for fluence monitoring. In addition to the fact that a linear relation exists between the nitrite formed and the applied fluence, nitrate is present in most natural waters and nitrite can be easily analyzed. Further research with respect to the effect of process parameters (water flow, choice of lamp, and total UV power) and water quality parameters needs to be performed.

Although they fulfill most of the criteria, vitamin C (ascorbic acid) and caffeine cannot be further explored for application as a fluence meter. The sensitivity of these compounds towards (visible) light and oxidation is so high that conversion already proceeds during normal handling. Because of this, no conversion was measured in collimated beam tests at fluences up to 1,000 J/m^2.

With respect to the absorption range, which is close to DNA, and the quantum yield (0.12 at 270 nm), the amino acid tryptophan is optional. Tryptophan can be analyzed with fluorescence. Moreover, there are indications that tryptophan can be found in natural water. However, a disadvantage of working with amino acids is the high risk of sample contamination.

The absorption spectrum of ozone is similar to that of DNA, and the quantum yield upon photolysis is about 0.9 (formation of $O(^1D)$), which makes ozone attractive. The application of ozone as a fluence meter would be based on detection of hydroxyl radicals formed. The enabling of an $^{\bullet}OH$ scavenger that reacts rapidly with $^{\bullet}OH$ must be dosed as well. Because this implies an indirect fluence measurement resulting in inaccurate fluence data, the ozone/scavenger system will probably be impracticable and expensive.

In order to decide whether uric acid and trans-urocanic acid can be developed as a fluence meter, data about the quantum yield (uric acid) and the options for (product) analysis (t-urocanic acid) are required.

In addition to using chemicals already present in the water as fluence meters, it is possible to add chemicals as, so to speak, external fluence meters. Chemicals that appear to fit the criteria outlined previously include the ferric ion (Fe^{3+}) and hydrogen peroxide with monitoring of the photochemically generated hydroxyl radical ($^{\bullet}OH$) using benzoate ion. These chemical systems will be described in more detail in the following paragraphs (see Table 2.1).

From the Kiwa (IJpelaar and Kruithof 2003) and Duke University (Sharpless, Cho, and Linden 2003) desk studies, including the preliminary tests, it was concluded that the research on development of a fluence meter would focus on nitrate/nitrite, Fe^{3+}/Fe^{2+} (ferric ion/ferrous ion), and the combination of hydrogen peroxide (H_2O_2) and benzoate ion.

The fact that nitrate is present in most natural waters and nitrite is relatively easily analyzed makes the use of nitrate attractive, especially in regard to the linear relation between nitrite formation and the fluence. The system is complicated because of the effect of several water quality parameters on nitrite formation.

Table 2.1
Overview of investigated natural and chemical compounds and final selection for further research

Compound	Presence in natural water	Potential	Selected
Fe^{3+}/Fe^{2+}	Yes*	Yes	Yes
H_2O_2/benzoate ion	No	Yes	Yes
Nitrate	Yes	Yes	Yes
Ozone	No	No	No
Tryptophan	Yes	yes	No
t-urocanic acid	No	No	No
Uric acid	Yes	No	No
Vitamin C	No	No	No

* Fe^{3+} concentration may be too low, indicating that iron may have to be added as well.

Both the Fe^{3+}/Fe^{2+} and the H_2O_2/benzoate ion systems enable the quantification of the fluence. Low concentrations of ferric iron (≤1 mg/L) are sufficient; the Fe^{2+} analysis is easy and can be applied online. However, to avoid significant flock formation, the iron system should be applied at pH <5. In contrast, the photo products formed using the H_2O_2/benzoate system are analyzed with high-performance liquid chromatography (HPLC), which makes this system less appropriate for online application. Nevertheless, this system is not restricted to a pH range.

CHAPTER 3
BACKGROUND

NITRATE ION PHOTOCHEMISTRY

Nitrate absorbs UV radiation between 200 and about 240 nm with a molar absorption coefficient of 9,900 $M^{-1}cm^{-1}$ at 200 nm (see Figure 3.1). During nitrate photolysis, nitrite is formed at a quantum yield of about 0.1.

Fundamental research on the photochemistry of nitrate ion at 254 nm has been performed by Mark et al. (1996). They found that the main products are nitrite (NO_2^-) and peroxynitrite (ONO_2^-/ONO_2H) for which the quantum yield depends on the pH (see Figure 3.2), nitrate content, oxidizability of organic compounds present, and the radiation intensity. Peroxynitrite was only detected in the alkaline range because the protonated form is unstable. Under alkaline conditions, a buildup of peroxynitrite may occur that may subsequently decompose into nitrite and nitrate. The main nitrate photolysis reactions are given here.

$$NO_3^- \xrightarrow{h\nu} NO_2^- + O(^3P) \qquad \Phi < 0.001 \qquad (3.1)$$

$$NO_3^- + H_2O \xrightarrow{h\nu} NO_2^\bullet + {}^\bullet OH + OH^- \qquad \Phi \approx 0.09 \qquad (3.2)$$

$$NO_3^- \xrightarrow{h\nu} ONO_2^- \qquad \Phi \approx 0.01 - 0.1 \qquad (3.3)$$

$$ONO_2H \longleftrightarrow ONO_2^- + H^+ \qquad pK_a = 6.6 \qquad (3.4)$$

$$ONO_2H \longrightarrow 0.7\ HNO_3 + 0.3\ {}^\bullet OH + 0.3\ NO_2^\bullet \qquad k = 0.8\ sec^{-1} \qquad (3.5)$$

Because of the low quantum yield compared to the Reactions 3.2–3.5, Reaction 3.1 can be neglected.

Nitrite formation appears to proceed primarily through the following fast reactions starting from Reaction 3.3 (Sharpless, Page, and Linden 2003). The degradation of peroxynitrite leads to the formation of the super oxide radical $O_2^{\bullet -}$ which reacts with $NO_2^\bullet$ forming peroxynitrate ($O_2NO_2^-$). This last species is believed to be important because the amount of nitrite produced through $O_2NO_2^-$ is directly linked to the $O_2^{\bullet -}$ concentration.

$$ONO_2^- \longrightarrow NO^\bullet + O_2^{\bullet -} \qquad k = 0.017\ sec^{-1} \qquad (3.6)$$

$$O_2^{\bullet -} + NO_2^\bullet \longrightarrow O_2NO_2^- \qquad k = 4.5 \cdot 10^9\ M^{-1}\ sec^{-1} \qquad (3.7)$$

$$O_2NO_2^- \longrightarrow NO_2^- + O_2 \qquad k = 1\ sec^{-1} \qquad (3.8)$$

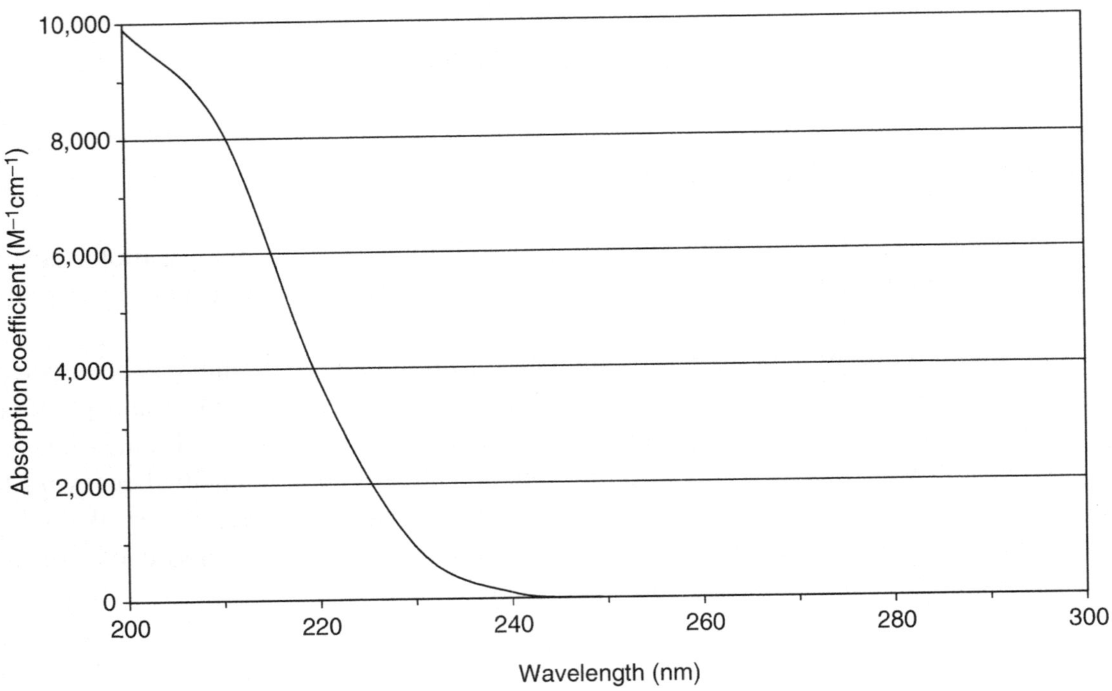

Figure 3.1 Absorbance of nitrate

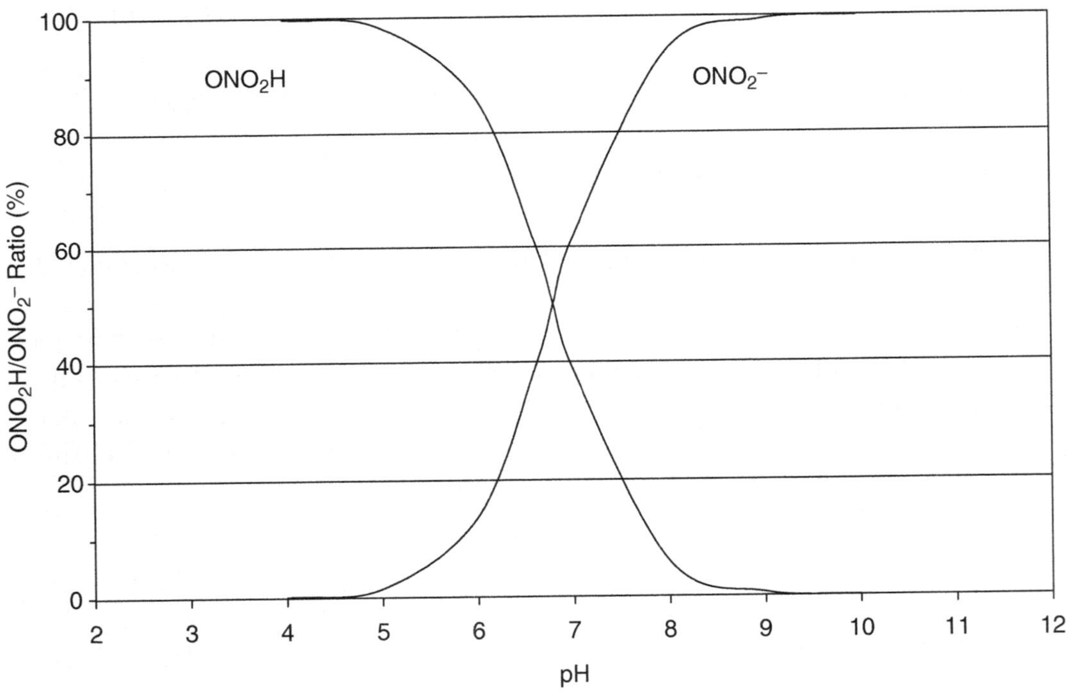

Figure 3.2 Equilibrium between ONO_2H and ONO_2^- versus pH

The reaction between the peroxynitrite ion ONO_2^- and radicals produced in natural water yields $ONO_2^{\bullet}$. The latter radical decays forms oxygen, and the nitrite radical $NO_2^{\bullet}$ finally leads to the formation of nitrite and nitrate:

$$ONO_2^- + {}^{\bullet}OH \longrightarrow ONO_2^{\bullet} + OH^- \qquad k = 4.5 \cdot 10^9\ M^{-1}sec^{-1} \qquad (3.9)$$

$$ONO_2H + {}^{\bullet}OH \longrightarrow ONO_2^{\bullet} + H_2O \qquad (3.10)$$

$$2\ ONO_2^{\bullet} \longrightarrow 2\ NO_2^{\bullet} + O_2 \qquad (3.11)$$

$$2\ NO_2^{\bullet} + H_2O \longrightarrow NO_2^- + NO_3^- + 2H^+ \qquad k = 4.7 \cdot 10^7\ M^{-1}sec^{-1} \qquad (3.12)$$

The nitrite ions can also react with hydroxyl radicals formed in Reaction 3.2:

$$NO_2^- + {}^{\bullet}OH \longrightarrow NO_2^{\bullet} + OH^- \qquad k = 10 \cdot 10^9\ M^{-1}sec^{-1} \qquad (3.13)$$

$$NO_2^{\bullet} + {}^{\bullet}OH \longrightarrow NO_3^- + H^+ \qquad (3.14)$$

According to Mark et al. (1996), Reaction 3.12 is one of the obvious pathways for the formation of nitrite through the $NO_2^{\bullet}$ radical.

Lymar and Hurst (1995, 1996) described the effects of carbon dioxide on the level of nitrite formation during irradiation of aqueous nitrate at 254 nm. They reported that carbon dioxide (CO_2) reacts rapidly with the peroxynitrite anion according to the following reaction:

$$ONO_2^- + CO_2 \longrightarrow ONO_2CO_2^- \qquad k = 2.9 \cdot 10^4\ M^{-1}sec^{-1} \qquad (3.15)$$

This second-order rate constant was established at pH ~6.2. Reaction rates of the order as calculated for Reaction 3.15 are comparable with the highest reaction rates reported for peroxynitrite with organics. The reaction between ONO_2H and HCO_3^- is much slower, probably due to the lower nucleophilic (ONO_2H) and electrophilic (HCO_3^-) characteristics of the reactants compared to ONO_2^- and CO_2, respectively. Whether the presence of dissolved CO_2 really affects the level of nitrite formation depends on its concentration with respect to the concentration of ${}^{\bullet}OH$ and the pH. The presence of CO_2 with respect to its equilibrium with bicarbonate ion (HCO_3^-) is strongly pH dependent. In the pH range typically applied for UV disinfection (e.g., 7.0–8.5), the relative appearance of CO_2 versus HCO_3^- is about 18% at pH 7 and only 1% at pH 8.5 (see Figure 3.3). Moreover, the reaction between ONO_2^- and CO_2 is orders of magnitude slower than the reaction between ONO_2^- and ${}^{\bullet}OH$. The reported reaction rate constants of these reactions are $2.9 \cdot 10^4\ M^{-1}sec^{-1}$ and $4.5 \cdot 10^9\ M^{-1}sec^{-1}$, respectively. This indicates that the concentrations of CO_2 and ${}^{\bullet}OH$ are most important regarding the faith of ONO_2^- towards nitrite formation. At low ${}^{\bullet}OH$ concentration, ONO_2^- may predominantly react with CO_2. The $ONO_2CO_2^-$ formed in Reaction 3.15 may decay according to the following reaction (Goldstein and Czapski 1998):

$$ONO_2CO_2^- \longrightarrow CO_3^{\bullet -} + NO_2^{\bullet} \qquad (3.16)$$

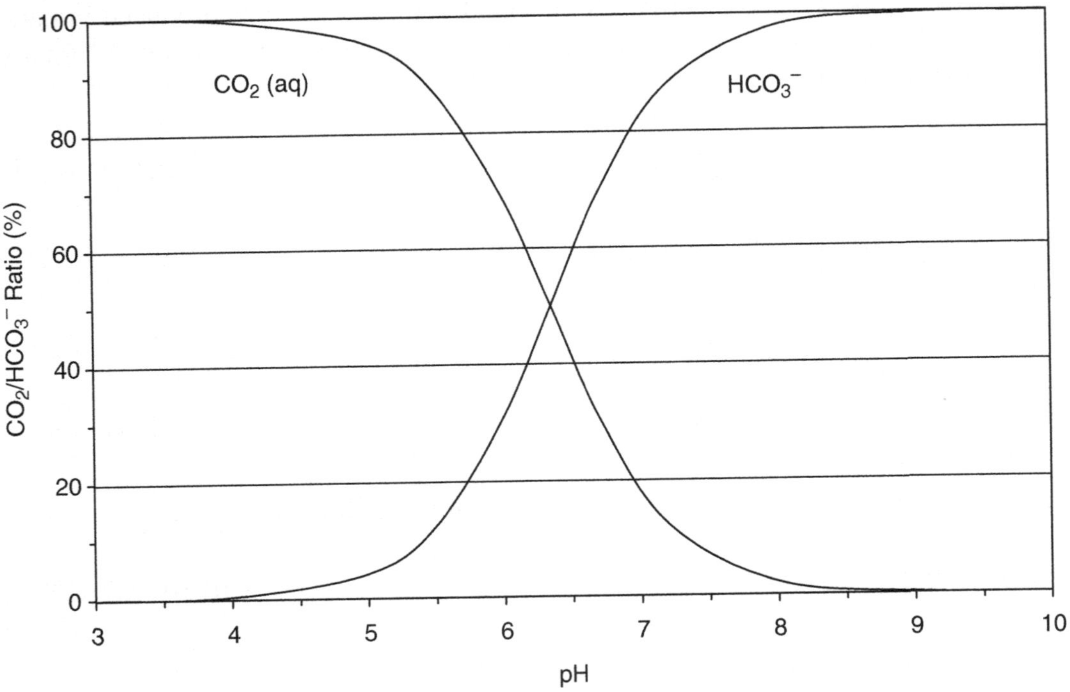

Figure 3.3 Equilibrium between CO_2 and HCO_3^- versus pH

Sharpless and Linden (2001) reported that dissolved inorganic carbon (dosed as $NaHCO_3$) significantly decreases nitrite formation at pH 8, but not at pH 6. They attributed this to the reaction between CO_2 and peroxynitrite. Lymar and Hurst (1996) measured the rate constant of the reaction between CO_2 and ONO_2^- at pH 6. Sharpless and Linden (2001) calculated the pseudo–first-order rate constant of Reaction 3.2 and found values for k' of 2–6 sec^{-1} (pH 6) and 0.6 sec^{-1} (pH 8). They also compared the rate of Reaction 3.2 with the rate Reaction 3.16, which shows the auto decomposition of peroxynitrite:

$$ONO_2^- \longrightarrow NO_3^- \qquad (3.17)$$

The apparent first-order rate constant for Reaction 3.16 was estimated giving rate constants of 1 and 0.08 sec^{-1} at pH 6 and 8, respectively. Since these findings agree with the results of earlier studies, Sharpless and Linden (2001) calculated that the reaction between ONO_2^- and CO_2 proceeds about 2–6 times (pH 6) and 10 times (pH 8) faster than the first-order decay process (Reaction 3.3) at these values.

Oxidizable organics may play a significant role in the nitrite ion formation process. Although the role of oxidizable organics in nitrite formation appears rather complex, it seems that these organics protect nitrite, peroxynitrite, and the $NO_2^{\bullet}$ radical from being oxidized to nitrate. Mark and co-workers (1996) summarized the most important reactions for the photochemical formation of nitrite from nitrate (see Figure 3.4). This suggests an important role of natural organic matter/dissolved organic carbon (NOM/DOC) which is ubiquitous in natural waters. The role of NOM on nitrite formation during nitrate photolysis has been studied in detail by Sharpless and Linden (2001). They found that in the presence of 5 ppm NOM (from the Suwannee River in

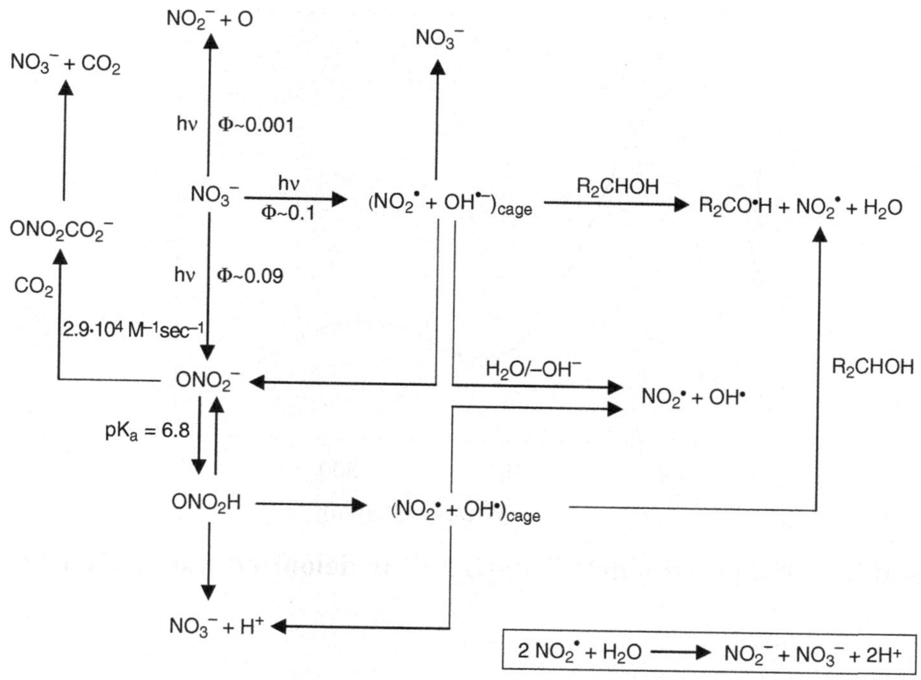

Source: Adapted from Mark et al. 1996 and Sharpless and Linden 2001.

Figure 3.4 Main reactions involved during nitrate photolysis in natural waters

the United States), the nitrite quantum yield was more or less constant with the fluence. However, the quantum yield of nitrite increased with the fluence in the presence of 15 ppm NOM. The increase in nitrite formation was explained by the scavenging of hydroxyl radicals ($^{\bullet}OH$) by NOM. In waters containing both CO_2 and DOC, nitrite formation may be increased by the reaction between organic compounds and $CO_3^{\bullet-}$.

Summarizing the effect of CO_2 and DOC on nitrite formation with respect to $^{\bullet}OH$ scavenging, the formation of nitrite may be increased according to

- Reaction of organic compounds with $CO_3^{\bullet-}$ allowing $NO_2^{\bullet}$ to form nitrite
- Reduction of $NO_2^{\bullet}$ to form nitrite directly

PHOTOCHEMISTRY OF THE Fe^{3+}/Fe^{2+} SYSTEM

Various hydroxy- and other ligated forms of ferric iron (Fe^{3+}) undergo photoreduction by charge transfer from the ligand according to reactions similar to Reaction 3.18.

$$FeOH^{2+} \xrightarrow{h\nu} Fe^{2+} + OH \qquad (3.18)$$

The monohydroxy complex shown in Reaction 3.1 is the predominant iron form in water at pH 4–5 and has a photolysis quantum yield of 0.15 at 313 nm (Faust 1994). The absorbance of Fe^{3+} in water extends below 200 nm, as shown in Figure 3.5.

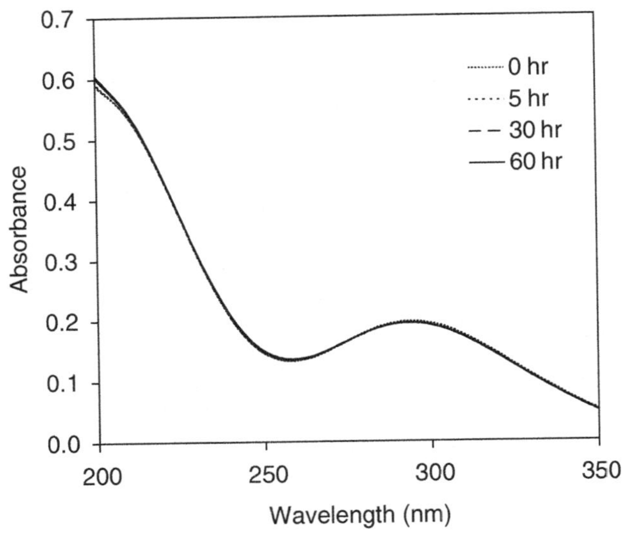

Figure 3.5 Stability of approximately 7 mg/L Fe^{3+} in deionized water, pH 4

Assuming that the quantum yield is constant down to 200 nm, this spectrum can be used to estimate Fe^{2+} formation rates as a function of germicidal UV fluence. Calculations using both low- and medium-pressure Hg lamp spectra show that the rate Fe^{2+} production in buffered water will be on the order of 4 to 6 nM m^2/J (40 to 60 nM cm^2/mJ) with 5 mg/L Fe^{3+} and approximately one fifth of this value at 1 mg/L. Details of this calculation are beyond the scope of this report (refer to Sharpless, Page, and Linden 2003 for details on how to model photodegradation kinetics).

Using the phenanthroline complexation method for measuring Fe^{2+} (APHA, AWWA, and WEF 1998), concentrations as low as approximately 0.9 μM can be accurately determined using a 5-cm cuvette. This implies that the Fe^{3+}/Fe^{2+} system may possibly be useful to quantify germicidal fluences as low as 200 J/m^2 using 5 mg/L Fe^{3+} (1,000 J/m^2 using 1 mg/L). If the Fe^{2+} quantum yield increases at shorter wavelengths, it may be possible to measure even lower fluence values.

PHOTOCHEMISTRY OF THE H_2O_2/BENZOATE SYSTEM

Although it is possible to use H_2O_2 directly as a fluence meter, its very low UV absorption makes its degradation kinetics too slow to be effective at germicidal irradiance levels. However, the concentration of hydroxyl radical that may be generated during UV/H_2O_2 treatment is high enough to detect its presence during the time required to deliver a germicidal fluence (Sharpless, Page, and Linden 2003). Photolysis of H_2O_2 in water generates hydroxyl radicals ($^{\bullet}OH$) with a quantum yield of 1 (Baxendale and Wilson 1957) according to

$$H_2O_2 \xrightarrow{h\nu} 2\,{}^{\bullet}OH \tag{3.19}$$

A simple method for monitoring $^{\bullet}OH$ is to follow the kinetics of reaction of a chemical probe with a known rate constant for reaction with $^{\bullet}OH$. At germicidal fluence levels, such reactions would need to produce nanomolar to micromolar changes in concentrations of affected chemicals to be useful as the basis for a fluence meter. The only reported probe that is both sensitive enough and has no known toxicity issues is the benzoate anion (used in soft drinks as a preservative).

Benzoate is a known and proven $^{\bullet}OH$ scavenger (Zhou and Mopper 1990) whose rate constant for reaction with $^{\bullet}OH$ is approximately $6 \cdot 10^9$ $M^{-1}sec^{-1}$ (Notre Dame Radiation Laboratory, University of Notre Dame, Indiana). The main products of this reaction are 2-, 3-, and 4-hydroxybenzoate, and these can be easily monitored by HPLC with UV detection (Zhou and Mopper 1990). Of the three products, 4-hydroxybenzoic acid provides the best analytical sensitivity because it has the most intense UV absorbance.

Assuming benzoate is the only $^{\bullet}OH$ scavenger in a water source, model calculations indicate that at 50 µM benzoate, the rate of 4-hydroxybenzoate formation will be approximately 5 and 30 nM m^2/J at H_2O_2 levels of 1 to 10 mg/L (29 to 290 µM), respectively. Again, details of these types of calculations can be found elsewhere (Sharpless and Linden 2003). With a quantification limit of 24 nM for 4-hydroxybenzoic acid, this implies that germicidal fluences lower than 10 J/m^2 can be accurately quantified in deionized (DI) water. In natural waters, $^{\bullet}OH$ scavenging by natural water constituents will increase the lowest fluence that can be quantified, and this effect can be anticipated in waters with lower UV transmittance.

UV DOSE MONITORING VERSUS REDUCTION EQUIVALENT FLUENCE

A fluence meter for UV disinfection systems should enable a fluence calculation with respect to the decimal inactivation of the microorganisms involved. This requires an accurate description of the relationship between the formation level of the photo product and the REF of the UV system. Preferably, compounds that absorb in the same wavelength range as that of the natural water disinfection part of the DNA (~ 240–280 nm) should be applied; however, this is not the case for the compounds studied here. Nitrate and iron (Fe^{3+}) absorb UV light predominantly below 240 nm. Therefore, the relationship between the photo product formation level and the REF should include a wavelength-dependent UV water absorbance correction. This is discussed in chapters 5 and 6 for the nitrate–nitrite system.

This difference in absorption characteristics between the photo chemicals studied here and DNA may cause a discrepancy between the "chemical" fluence and the REF. This may be the case with regard to the following situations:

- Compounds that absorb UV radiation in the same wavelength range as nitrate affecting the nitrite formation but leaving the disinfection capacity of the UV system unchanged
- Compounds that absorb UV radiation in the same range as DNA affecting the disinfection capacity but leaving the level of nitrite formation unchanged
- Compounds affecting both the disinfection capacity and the amount of nitrite formed but in a different way

Whether nitrate and iron will result in a UV disinfection monitoring tool will be discussed in the following chapters. Alternatively, a so-called alarm system may be developed to warn end users when the chemical fluence is outside the acceptable fluence range.

SUMMARY

Nitrate Ion Photochemistry

The literature survey described in this report dealt with the water quality parameters affecting the photolysis of nitrate and the quantum yield (~0.1) of nitrite formation. Besides the initial nitrate content $[NO_3^-]_0$ of the water, two types of compounds should be distinguished:

1. Compounds that physically interfere in the nitrite formation process, e.g., $UVT_{200–240}$
2. Compounds that chemically interfere in the nitrite formation process by changing the nitrite quantum yield, i.e., oxidizable organics (DOC) and CO_2 in equilibrium with HCO_3^-

The CO_2/HCO_3^- equilibrium as well as the ONO_2H/ONO_2^- equilibrium is represented by changes in pH and as such should be applied in the proposed monitoring system.

It can be concluded that nitrite formation proceeds essentially according to the products of Reactions 3.2, 3.3, 3.6–3.8, and 3.12. The most important species are ONO_2^-/ONO_2H, $NO_2^{\bullet}$, $O_2^{\bullet-}$, $^{\bullet}OH$, and $O_2NO_2^-$. The formation of peroxynitrite is essential because it leads to nitrite by two different pathways:

- Through reaction with hydroxyl radicals ($^{\bullet}OH$)
- By reaction with superoxide radicals ($O_2^{\bullet-}$), leading to the formation of peroxynitrate and finally nitrite

In chapter 4, the effect of the above-described parameters on nitrite formation is quantified. Subsequently, based on the extent of the effects, parameters are selected to include in a practical fluence monitoring system. A proposed system is described in chapter 6.

Photochemistry of the Fe^{3+}/Fe^{2+} System

UV irradiation of ferric iron (Fe^{3+}) leads to the formation of ferrous iron (Fe^{2+}, quantum yield about 0.15 at 313 nm), which can be detected by complexation with phenanthroline. Using this method, ferrous iron concentrations as low as 0.9 μM (~50 μg/L Fe^{2+}) can be detected. Germicidal fluences of 200 J/m^2 may possibly be quantified applying a 5-mg/L Fe^{3+} dose. Even lower fluences may be calculated if the Fe^{2+} quantum yield increases at shorter wavelengths.

Photochemistry of the H_2O_2/Benzoate System

The H_2O_2/benzoate system is based on the UV photolysis of H_2O_2 with a quantum yield of about 1. The subsequently formed $^{\bullet}OH$ are scavenged by benzoate ion forming 2-, 3-, and 4-hydroxybenzoate. The high UV absorbance of 4-hydroxybenzoate enables this product to provide the lowest detection limit. This implies that germicidal fluences as low as 10 J/m^2 can be accurately calculated in DI water. However, the lowest fluence to be quantified will increase using natural waters due to the presence of natural $^{\bullet}OH$ scavengers such as NOM/DOC and alkalinity.

CHAPTER 4
MATERIALS AND METHODS

OVERVIEW

All UV tests in this research were carried out with a collimated beam apparatus (see Figure 4.1). Irradiation times were measured using an automatic shutter device. The UV fluence was determined using the spreadsheet developed by Bolton and Linden (2003; see also the International Ultraviolet Association [IUVA] Web site, www.iuva.org). Throughout this study, the fluences were corrected for DNA absorption (for the germicidal action curve, see Meulemans 1987) based on which fluences up to 1,000 J/m^2 (nitrate) and 2,000 J/m^2 (Fe^{3+}/Fe^{2+} and H_2O_2/benzoate) were applied.

NITRATE ION

Experimental Overview

The collimated beam apparatus was equipped with a 1.5-kW medium-pressure UV lamp (operated at 50%) from Philips Lighting BV (Roosendaal, the Netherlands). Water samples of about 60 mL were irradiated in a petri dish, which was located at a fixed distance from the UV lamp. The intensity of the UV lamp was measured with an IL 1700 Research Radiometer and SED240W detector, both from International Light (Newburyport, Mass.). The irradiation times were determined using the protocol developed by Bolton and Linden (2003). The spreadsheet available from the IUVA Web site was used to calculate the required irradiation times for the desired fluences.

In all cases, the water quality applied was IJssel Lake water pretreated by coagulation (ferric iron), sedimentation, and rapid sand filtration by PWN Water Supply Company North Holland (see Table 4.1). The water was used as received, and the tests were carried out at the same day. Table 4.1 shows the range of the parameters used in this study. The water quality parameters assigned during the tests are indicated with the figures in chapter 5.

The nitrite concentrations measured after irradiation in the collimated beam tests were corrected for the nitrite levels in the IJssel Lake water.

Tests and Reagents

Tests that studied the effect of the initial nitrate content of the water were carried out by dosing sodium nitrate (Chemical Abstracts Service [CAS] No. 7731-99-4, American Chemical Society/U.S. Pharmacopeia [ACS/USP], J.T. Baker distributed by Fisher Scientific BV's Hertogenbosch, the Netherlands) to the pretreated IJssel Lake water. The final concentrations of the samples ranged between 3.5 and 15 mg/L NO_3^-.

In order to measure the effect of DOC on the nitrite formation, DOC concentrations (measured as NPOC) between 2 and 13 mg C/L were obtained by concentrating the initial IJssel Lake DOC level using nanofiltration. As a result, the initial nitrate concentration increased as well but was kept in between the maximum concentrations encountered in the IJssel Lake water.

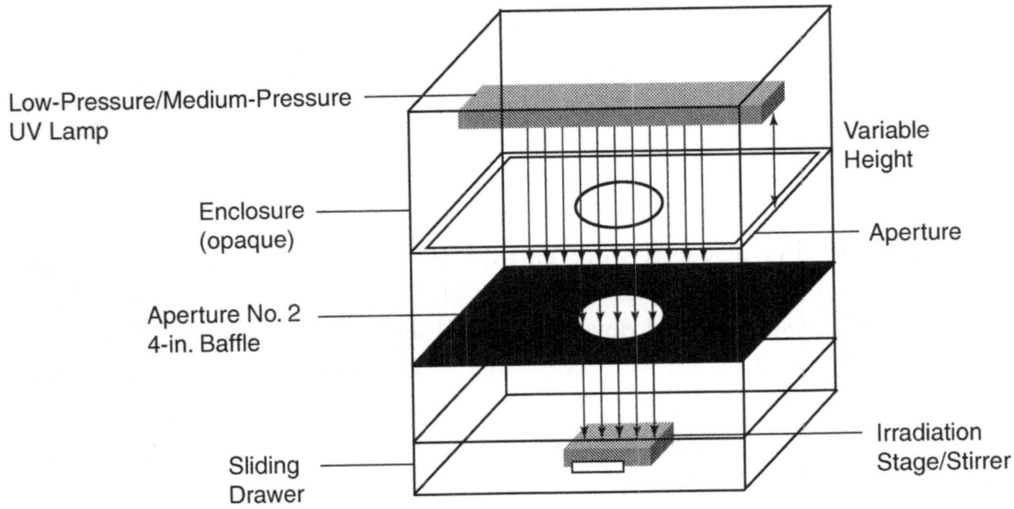

Figure 4.1 Schematic presentation of the collimated beam device applied throughout this research

Table 4.1
Water quality parameters of pretreated IJssel Lake water

Date	$[NO_3^-]_0$ (mg NO_3^-/L)	pH	Alkalinity (mg HCO_3^-/L)	NPOC* (mg C/L)	$[NO_2^-]_0$ (µg NO_2^-/L)
10-29-03	3.5	7.87	113	3.0	4.5
11-27-03	3.5	7.88	119	3.0	4.0
12-17-03	3.9	7.78	121	3.5	6.1
01-22-04	5.7	7.75	125	2.5	7.9
01-30-04	6.6	7.77	131	3.5	7.5
02-26-04	11.2	7.80	136	3.6	16.8
03-10-04	12.1	7.52	133	3.6	15.8
03-25-04	13.1	7.38	131	3.7	13.1
04-14-04	12.3	7.70	138	4.1	3.4
04-27-04	11.5	7.78	145	4.0	8.3
05-26-04	10.0	7.61	145	4.1	3.0
06-30-04	6.2	7.52	125	4.4	3.4
07-27-04	3.2	7.54	120	5.5	5.6
08-31-04	2.9	7.89	110	4.6	1.4

* NPOC: Nonpurgeable organic carbon.

Tests with variation in bicarbonate content were performed by adding HCO_3^- realizing concentrations of 136, 171, 204, and 238 mg/L. To that end, sodium bicarbonate (CAS No. 144-55-8, ACS/USP, J.T. Baker) was applied.

Analysis Methods

Nitrite was determined by a spectrophotometric method (conforms to Netherlands Standards Institute [NEN] 6474; NEN 1981b) at a wavelength of 542 nm. The detection limit using a 10-cm cuvette is 0.5 μg N/L.

Nitrate was determined by ion chromatography (USEPA 1991) with a detection limit of 10 μg N/L.

NPOC was measured on a Shimadzu TOC 5000A apparatus (Shimadzu Benelux, Hertogenbosch, the Netherlands). The inorganic carbon is removed by acidification of the sample with hydrochloric acid (HCl). Subsequently, carbon-free air is used to purge the CO_2 formed from the solution. The organic carbon is converted to CO_2 at 680°C over a platinum catalyst. The detection limit for this method is 0.2 mg C/L (MilliQ water; Millipore gradient A10, Amsterdam, the Netherlands).

Bicarbonate was determined by titration with 0.02-M HCl and the solution was titrated down to a pH of 8.35. This method has a detection limit of 6.1 mg HCO_3^-/L.

The pH was determined by a potentiometric method (conforms to NEN 6411, 1981a).

THE Fe^{3+}/Fe^{2+} AND H_2O_2/BENZOATE SYSTEMS

Experimental Overview

Two sets of experiments were performed: one to investigate Fe^{2+} formation from Fe^{3+} photolysis and another to investigate 4-hydroxybenzoate formation in solutions of sodium benzoate and H_2O_2. All experiments were performed in duplicate, and the results presented here are averages and standard deviations of the two experiments. Both low-pressure (LP) and medium-pressure (MP) Hg lamps in quasi-collimated beam apparatuses were used for all experiments, and all experiments were conducted with a solution depth of 0.9 cm to minimize any potential artifacts due to divergence of the beam during passage through the solution (this ensures that the calculated UV fluence rates are accurate). Experiments with both systems were carried out in DI water and synthetic surface water prepared in the lab. Additional experiments with Fe^{3+} were conducted in water obtained from drinking water sources in North Carolina and New York.

Reagents

All chemicals were analytical or USP grade and were used as received. For the Fe^{3+} stock solution, 0.242 g of $FeCl_3 \cdot 6H_2O$ was dissolved in 0.01 N HCl resulting in a concentration of 500 mg/L Fe^{3+} at pH 2. A 54-mM benzoic acid stock solution was prepared by adding 0.65 g benzoic acid and approximately 0.2 g of sodium hydroxide (NaOH) to 100 mL DI water and then was stirred overnight to achieve complete dissolution. Hydrogen peroxide (H_2O_2) was added to solutions by diluting 30% H_2O_2 (unstabilized, Ultrex brand, J.T. Baker), which by analysis was shown to be 10 M (340 g/L).

Table 4.2
Water quality of natural waters used for Fe^{3+} experiments

Source water	$[Fe^{3+}]$ (mg/L)	TOC* (mg/L)	Alkalinity as $CaCO_3$	Hardness as $CaCO_3$	Nitrate (mg N/L)
Durham, N.C. (3/04)	0.5	9.1	17	44	0.67
Durham, N.C. (6/04)	ND†	10.9	19	63	0.53
NYC-579, pH 6	ND	16.7	55.5	175	0.53
NYC-570, pH 6	ND	3.2	7	16	0.47

* TOC = total organic carbon.
† ND = not determined.

Experimental Waters

All pH adjustments were made with either 1 N HCl or 1 N NaOH. Experiments were conducted in DI water whose pH was adjusted to 4 (Fe^{3+} experiments), 1 mM phosphate buffer at pH 7 (benzoic acid experiments), and in a synthetic surface water (both Fe^{3+} and benzoic acid experiments). Phosphate buffer was prepared by diluting a 1 M stock solution at pH 7. The pH of the diluted solutions was adjusted to either 4 (for iron experiments) or 7 (for benzoic acid experiments). Synthetic surface water was prepared at pH 4 and pH 8 for iron experiments and benzoic acid experiments, respectively. Two types of NOM were used to prepare the synthetic water: Suwannee River NOM (International Humic Substances Society, University of Minnesota, St. Paul, Minn.) and alginic acid (Sigma-Aldrich, St. Louis, Mo.). The synthetic water at both pH 4 and 8 contains several inorganic salts ($CaCl_2$, $MgCl_2$, $NaNO_3$, and $CaSO_4$) and NOM. At pH 8, the water also contains 1.5 mM $NaHCO_3$, and a small amount (approximately 2 mM) of phosphate buffer. At pH 4, neither $NaHCO_3$ nor phosphate buffer was added because the solution is well buffered by H^+, and the bicarbonate would all be converted to CO_2 upon adjusting the pH.

Three natural waters were also used for studying Fe^{3+}. These are source waters obtained from the Williams Water Treatment Plant in Durham, N.C., and from the Cat-Del water supply in New York (two samples from unknown locations in the supply system). All waters were analyzed for Fe^{3+} and other water quality parameters. In one instance, Durham water was found to contain 0.5 mg/L Fe^{3+}, but Fe^{3+} was undetectable in both New York City (NYC) water samples. The major water quality parameters for the waters are given in Table 4.2.

Photolysis Experiments

All experiments were carried out with water in equilibrium with respect to O_2 in the atmosphere. In each experiment, 20-mL samples were placed in 53-mm-diameter petri dishes (solution depth = 0.91 cm) and irradiated using bench-scale UV quasi-collimated beam LP or MP systems. Germicidal UV fluences ranging from 0 to 2,000 J/m^2 were used for all experiments. The fluence was calculated as the exposure time multiplied by the germicidal irradiance, which was determined using a spreadsheet calculation from radiometer readings, solution absorbance, sample depth, and the lamp spectrum weighted by the absorption spectrum of DNA; for LP experiments, no germicidal weighting factor was applied (i.e., the average irradiance was assumed to be the germicidal irradiance). During exposure, the samples were stirred slowly at room temperature. After predetermined exposure times to achieve a specific germicidal fluence, samples were removed from the dish and analyzed for the relevant chemical species.

Fe^{2+} Analysis

Fe^{2+} was determined using the phenanthroline method (APHA, AWWA, and WEF 1998) modified by the introduction of a cuvette with a 5-cm optical path length to measure low absorbance values. Analyses were performed using 5-mL samples that were mixed with 20 mL of colorimetric reagent (10 mL of phenanthroline solution, 7.5 mL of acetate buffer solution, and 2.5 mL DI water). After 45 minutes, the Fe^{2+} was calculated from the absorbance at 510 nm based on the calibration curve, which was constructed using Fe^{2+} solutions at pH 4 of 0.3, 0.6, 1, 2, and 5 µM. Using this approach, the Fe^{2+} quantification limit was approximately 0.20 µM in DI water and 0.84 µM in synthetic surface water.

H_2O_2 Analysis

Hydrogen peroxide was determined by the I_3^- method (Klassen, Marchington, and McGowan 1994). This method uses the absorbance at 352 nm of I_3^- formed by the reaction of H_2O_2 and I^- to determine the original H_2O_2 concentration using the relationship $\varepsilon_{352} = 26{,}400\ M^{-1}cm^{-1}$. For all measurements, the absorbance of a blank (no H_2O_2, analyzed identically to samples) was subtracted from sample readings.

Hydroxybenzoic Acid Analysis

Reversed-phase HPLC was used for analysis of benzoic acid, 4-hydroxybenzoic acid, and 3-hydroxybenzoic acid. After photolysis experiments in which the acids were present as the benzoate conjugate bases, 0.3-mL samples were acidified with 10 µL of 0.15 M H_3PO_4 to pH ~2, to convert them to the acid form, and then filtered (0.45 µm) prior to analysis. Separation was achieved with a C18 reverse-phase column (Altech Prevail [Alltech Associates Inc., Deerfield, Ill.], 3 µm particle size, 53 mm × 7 mm) and a mobile phase of 1:3 acetonitrile:water (pH 2.5) at a flow rate of 1.5 mL/min. Analytes were detected by absorbance from 245 to 255 nm; and with 50-µL injections, the estimated quantitation limit for 4-hydroxybenzoic acid (the preferred analyte due to high UV absorption) is 24 nM in DI water.

CHAPTER 5
RESULTS AND DISCUSSION

NATURALLY OCCURRING COMPOUNDS

Nitrate Ion

Introduction

To be able to use the level of nitrite formation as a measure of the fluence delivered to the water, the operating conditions (water flow and total UV input) as well as the water quality parameters ($[NO_3^-]_0$, pH, DOC, and $UV_{200–240}$) must be known.

Figure 5.1 shows a nitrite–fluence line for the pretreated surface water applied in this research. It shows a clear linear relationship between nitrite formation and fluence in a typical natural water source. The level of nitrite produced here is the result of the sum of the contributions of the water quality parameters as discussed in chapter 3.

Figure 5.2 clearly shows that the value of the water quality parameters affecting the nitrite formation varies significantly over time. To enable monitoring of the fluence delivered to the water in practical UV systems, the contribution of each of the above-described parameters on nitrite formation must be quantified. The question is: to what extent do the relevant water quality parameters affect the nitrite formation? Based on the importance of each parameter, a correction of the nitrite data can be pursued by determining the fluence from the nitrite formation under all possible conditions.

UV Transmission ($UVT_{200–300}$)

One of the water quality parameters strongly affecting the efficacy of UV disinfection systems is the transmission of the water for UV radiation. In order to have an efficient and cost-effective treatment, both the radiation distribution over the volume of the reactor and the residence time distribution of the water in the reactor should be in a narrow range.

UV-absorbing compounds such as nitrate and organic carbon (DOC/TOC) have a significant effect on the penetration depth of UV light. This also counts for solid particles that may cause significant scattering of the radiation. However, the contribution of particles shielding the target microorganisms from being irradiated is not within the scope of this research. This aspect is included in the overall water transmission.

Regarding the application of a nitrate fluence meter, changes in absorbance in the 200–240-nm range directly affect the level of nitrite formation. In many natural waters, changes in absorbance are almost completely due to variation in the initial nitrate content ($[NO_3^-]_0$) of the water and are hardly caused by organic carbon. This is due to two aspects. First, according to the Beer-Lambert law, the absorbance is 1.79 at a nitrate concentration of 11.2 mg/L (absorption coefficient 9,900 $M^{-1}cm^{-1}$ at 200 nm, light path 1 cm). This is equal to the absorbance measured with the pretreated IJssel Lake water with the same nitrate content (see Figure 5.3). Secondly, the contribution of organic carbon to the absorbance in the 200–240-nm range is much lower than for nitrate (compare Figure 5.4 with Figure 5.5). In addition, the organic carbon content varies generally much less than the nitrate level.

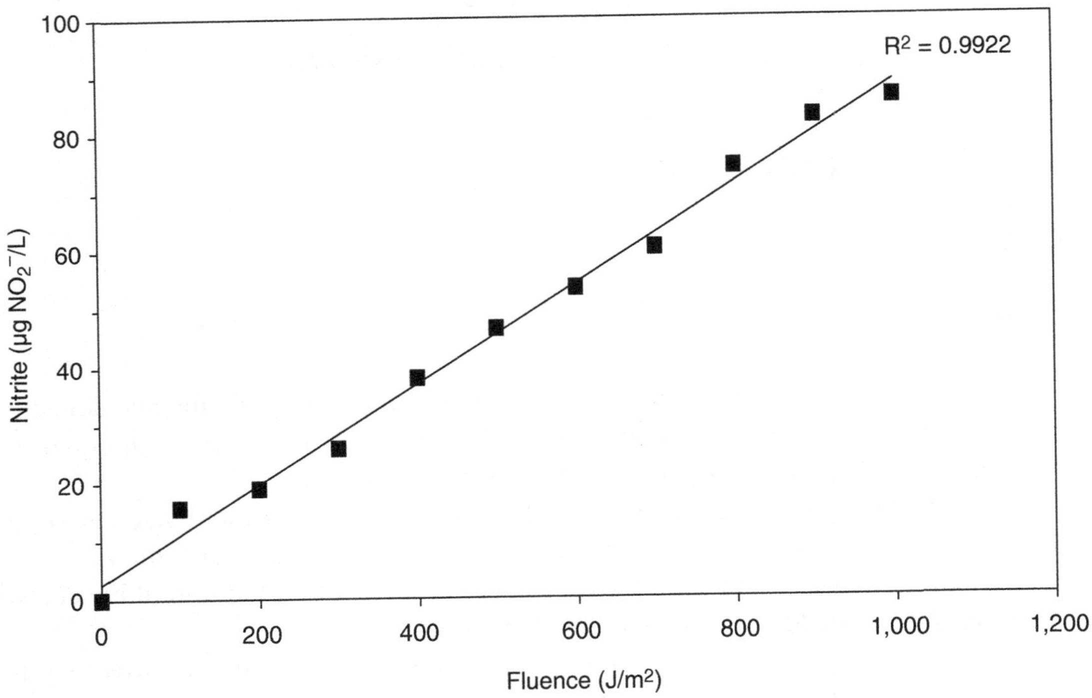

Figure 5.1 Typical relation between nitrite formation and the UV fluence ($[NO_3^-]_0$ = 14 mg/L, DOC = 3.8 mg/L, alkalinity = 150 mg HCO_3^-/L, pH = 7.5)

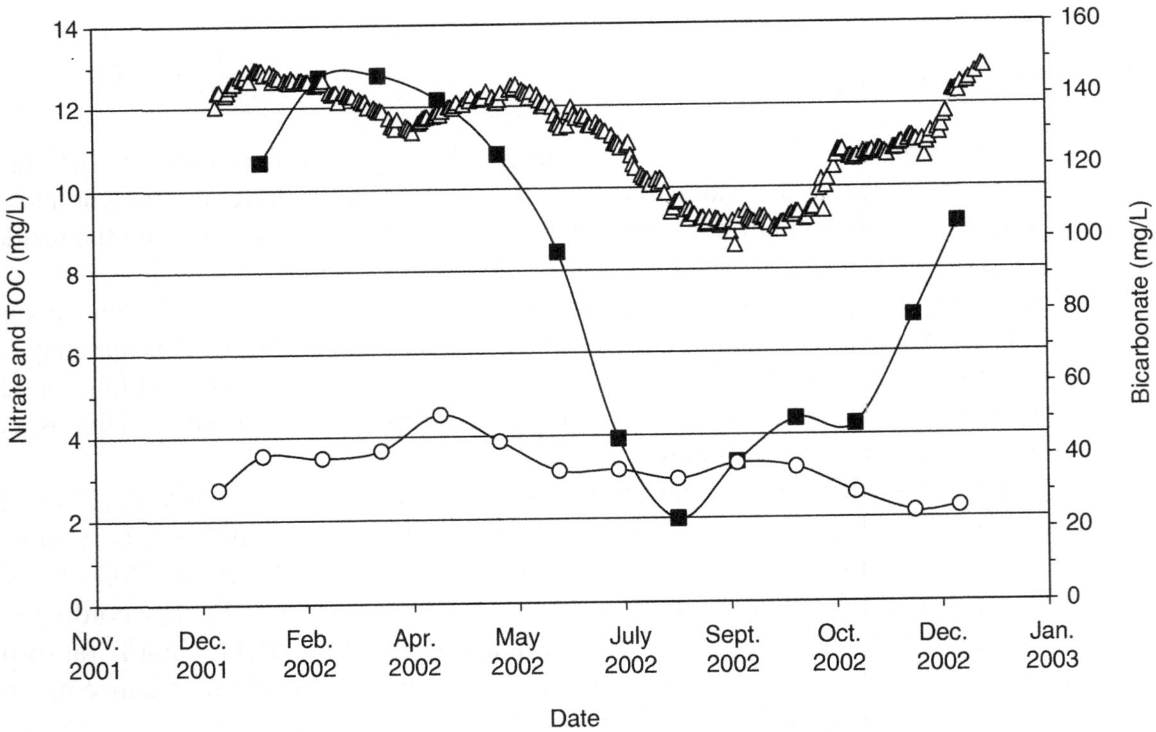

Figure 5.2 Variation in water quality of the pretreated IJssel Lake water recorded over time: NO_3^- (▪), TOC (○), HCO_3^- (△). See also Table 4.1.

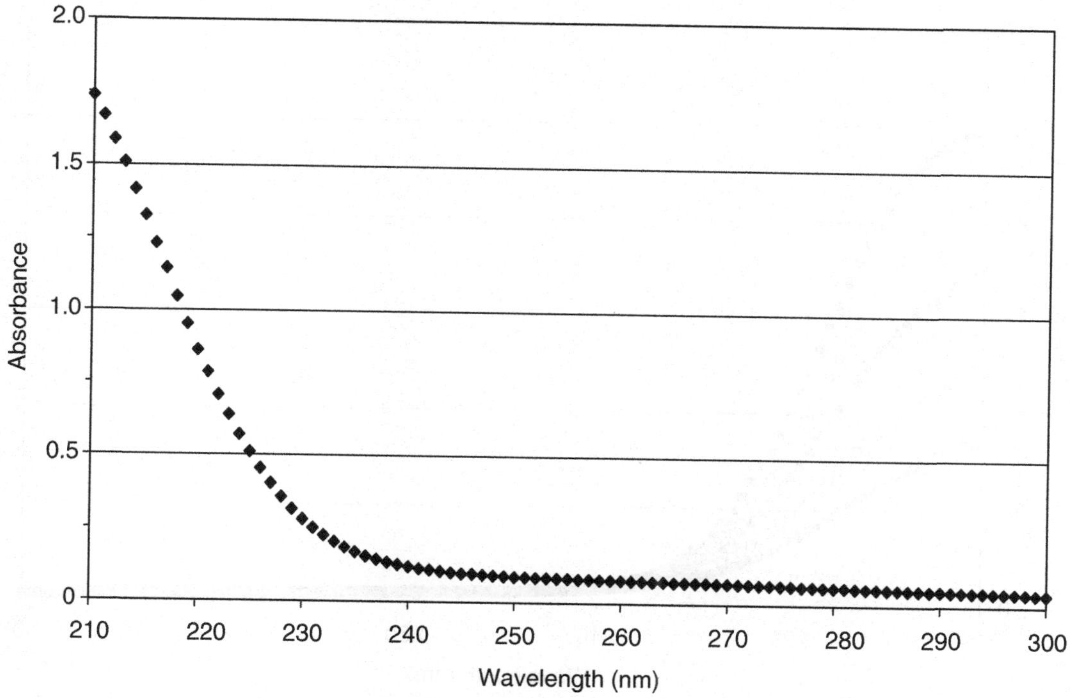

Figure 5.3 Absorbance of the pretreated IJssel Lake water. Water quality: 11.2 mg NO_3^-/L, NPOC = 3.6 mg C/L.

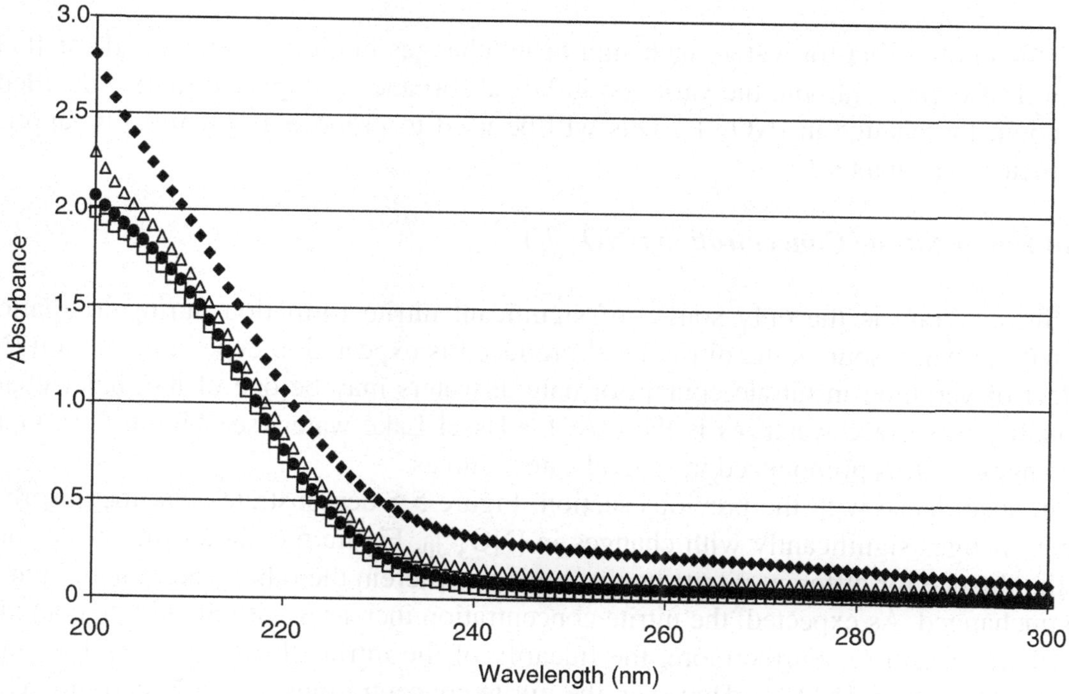

Figure 5.4 Change in absorbance of the pretreated IJssel Lake water with organic carbon content (mg C/L): 2.1 (□), 3.1 (●), 6.2 (△), 13.4 (◆); $[NO_3^-]_0$ = 10.3–10.9 mg/L

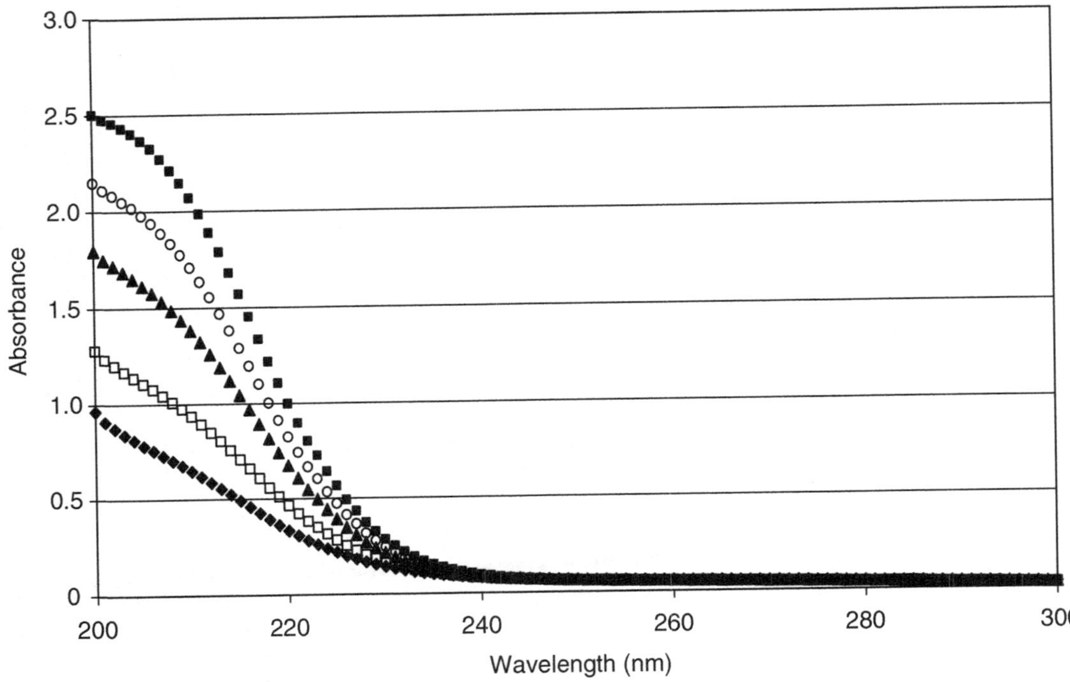

Figure 5.5 Change in absorbance of the pretreated surface water with nitrate content (mg NO_3^-/L) of 3.5 (♦), 5.9 (□), 9.4 (▲), 12.1 (○), 15.0 (■); pH = 7.87, alkalinity = 113 mg HCO_3^-/L, NPOC = 3 mg C/L

This implies that for waters with significant changes in nitrate content and small changes in the level of organic carbon, the variation in UV absorbance is expected to give detailed information about the change in $[NO_3^-]_0$. This will be used to propose a system to be applied as a fluence meter (see chapter 6).

Effect of Initial Nitrate Concentration ($[NO_3^-]_0$)

Since nitrate is the only source of significant nitrite formation during irradiation of a typical natural water source, the nitrite level produced is expected to depend largely on $[NO_3^-]_0$. The effect of variation in nitrate content of natural waters may be one of the major changes in water quality in surface water, as is the case for IJssel Lake water (see Figure 5.2). Generally, these changes are less pronounced in groundwater sources.

In accordance with the previous section, Figure 5.5 demonstrates that the absorbance of the water changes significantly with changes in $[NO_3^-]_0$. Figure 5.6 shows the nitrite formation versus the fluence for different nitrate concentrations, wherein the other water quality parameters remain unchanged. As expected, the nitrite concentration increases with the nitrate content of the irradiated water sample. Furthermore, the linearity of the nitrite–fluence lines is not negatively affected by changes in $[NO_3^-]_0$. However, the nitrite concentrations do not linearly increase with $[NO_3^-]_0$ but level off, as can be seen in Figure 5.7. This implies that upon changes in $[NO_3^-]_0$ the contribution of (an) other water quality parameter(s) contributing to nitrite formation may become apparent. This may be due to the level of oxidizable organic compounds present in the water. Elevated levels of nitrate will increase the number of nitrite radicals ($NO_2^•$) (Reaction 3.2). At a

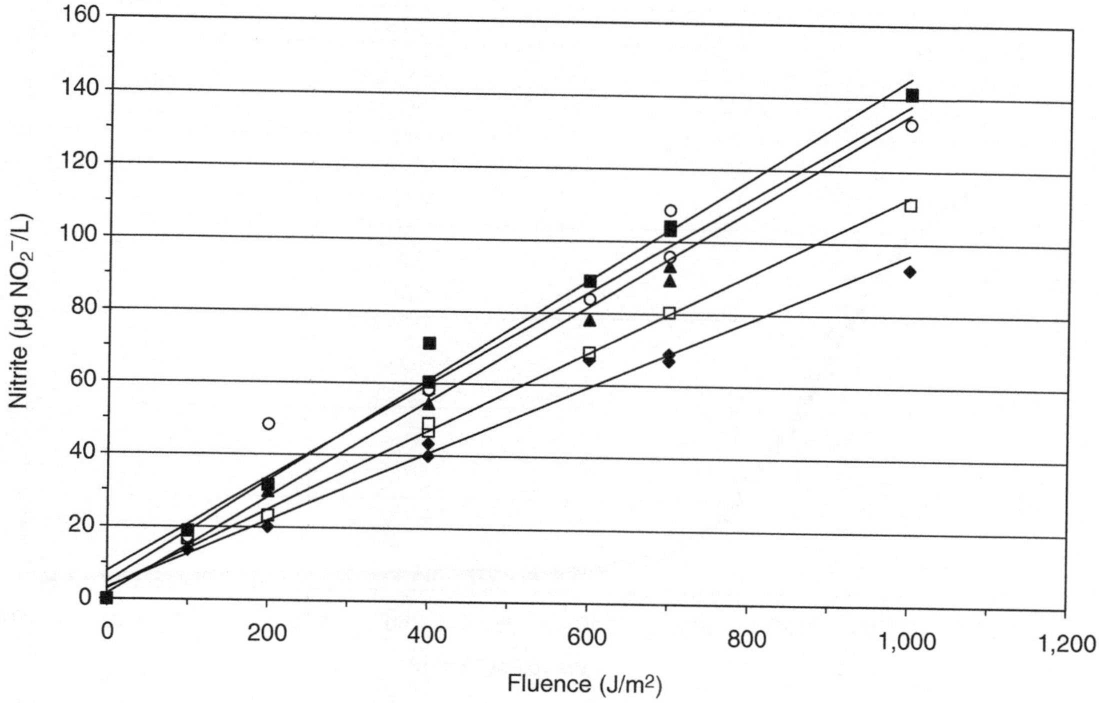

Figure 5.6 Nitrite formation versus the UV fluence for nitrate concentrations (mg NO_3^-/L) of 3.5 (◆), 5.9 (□), 9.4 (▲), 12.1 (○), 15.0 (■). Water quality: pH = 7.87, alkalinity = 113 mg HCO_3^-/L, NPOC = 3 mg C/L.

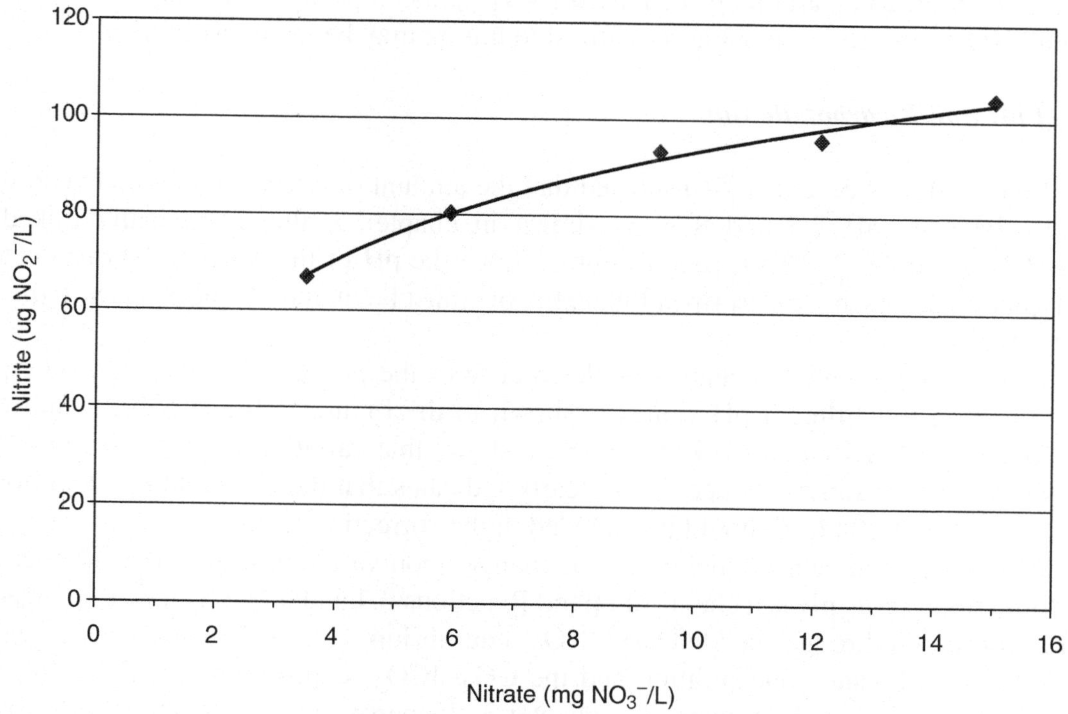

Figure 5.7 Nitrite formation versus initial nitrate content at 400 J/m²

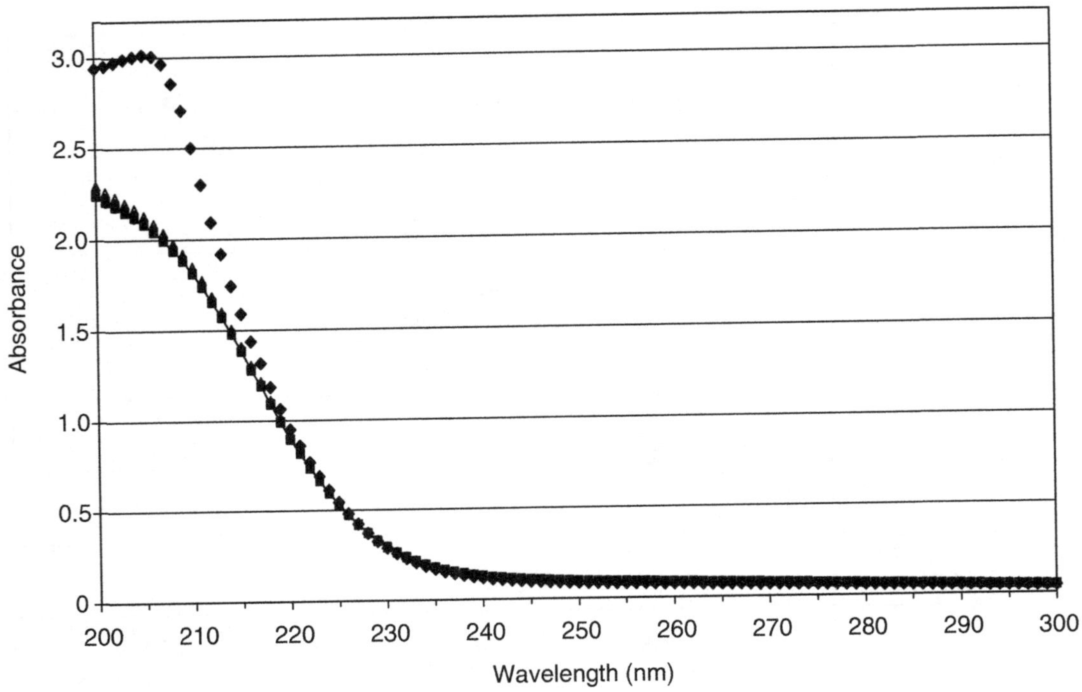

Figure 5.8 Change in absorbance of the pretreated IJssel Lake water with pH 6.50 (○), 7.00 (▲), 7.70 (□), 8.52 (●), 9.05 (△), 11.93 (◆). Water quality: 12.3 mg NO_3^-/L, alkalinity = 138 mg HCO_3^-/L, NPOC = 4.1 mg C/L.

distinct level of organic carbon but increased $[NO_3^-]_0$, the capacity of the oxidizable organics to protect all $NO_2^•$ radicals from being reoxidized to nitrate may become insufficient.

Effect of pH and Bicarbonate Ion

Mark and co-workers (1996) showed that the amount of nitrite formed increases with pH, especially between pH 7.5 and 8.5. As shown in chapter 3, this is basically linked to the ONO_2H/ONO_2^- and CO_2/HCO_3^- equilibriums. Since the pH of the water to be disinfected with UV is typically in the above-described range, the pH must be included with respect to the level of nitrite formation.

Figure 5.8 presents the change in absorbance of the pretreated water versus the pH. The nitrite formation for different pH values is shown in the Figures 5.9 and 5.10. Confirming the results found by Mark and co-workers (1996), it shows that variation in pH in the 7.0–8.5 range affects the amount of nitrite formed. This clearly indicates that the pH should be monitored, and its effect on nitrite formation should be included in the correction factor.

It shows that the absorbance does not change upon variation in pH. This implies that the nitrite formation versus pH is a chemical effect. Reactions 3.4 and 3.6–3.8 show the impact of pH on nitrite formation through the ONO_2H/ONO_2^- equilibrium (see also Figure 3.2). However, the effect of the bicarbonate concentration and the CO_2/HCO_3^- equilibrium on nitrite formation is more difficult to indicate. As explained in chapter 3, the peroxynitrite ion ONO_2^- is an important intermediate in one of the nitrite formation pathways. It was also shown that the presence of CO_2 may react with ONO_2^-, decreasing its concentration (see Figure 3.4).

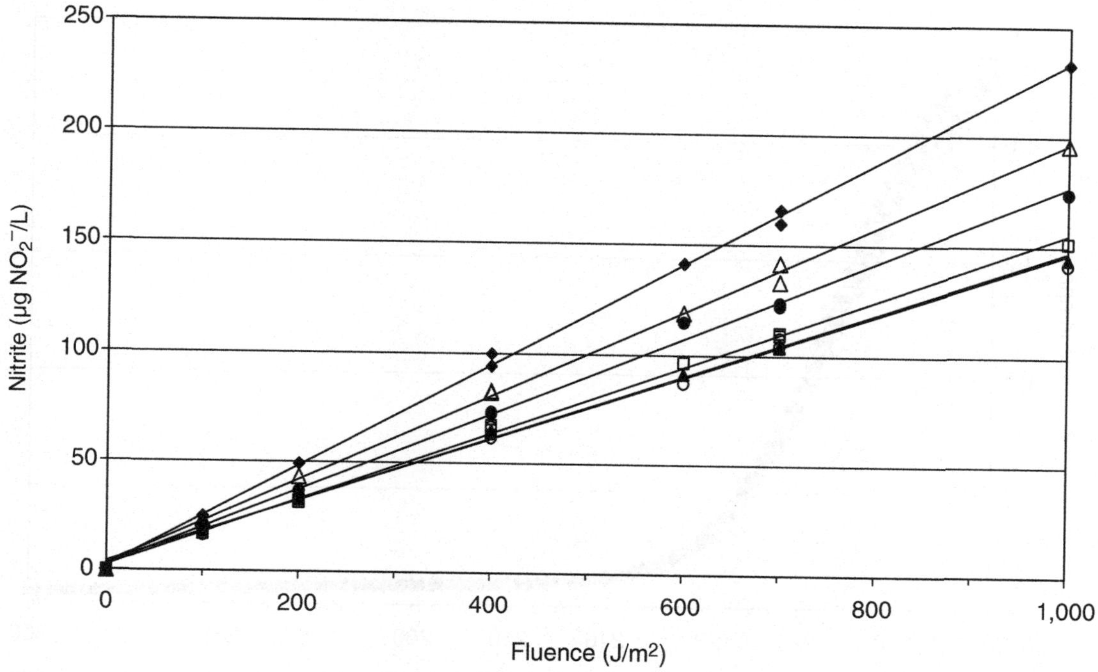

Figure 5.9 Nitrite formation versus the UV fluence for pH values 6.50 (○), 7.00 (▲), 7.70 (□), 8.52 (●), 9.05 (△), 11.93 (◆). Water quality: 12.3 mg NO_3^-/L, alkalinity = 138 mg HCO_3^-/L, NPOC = 4.1 mg C/L.

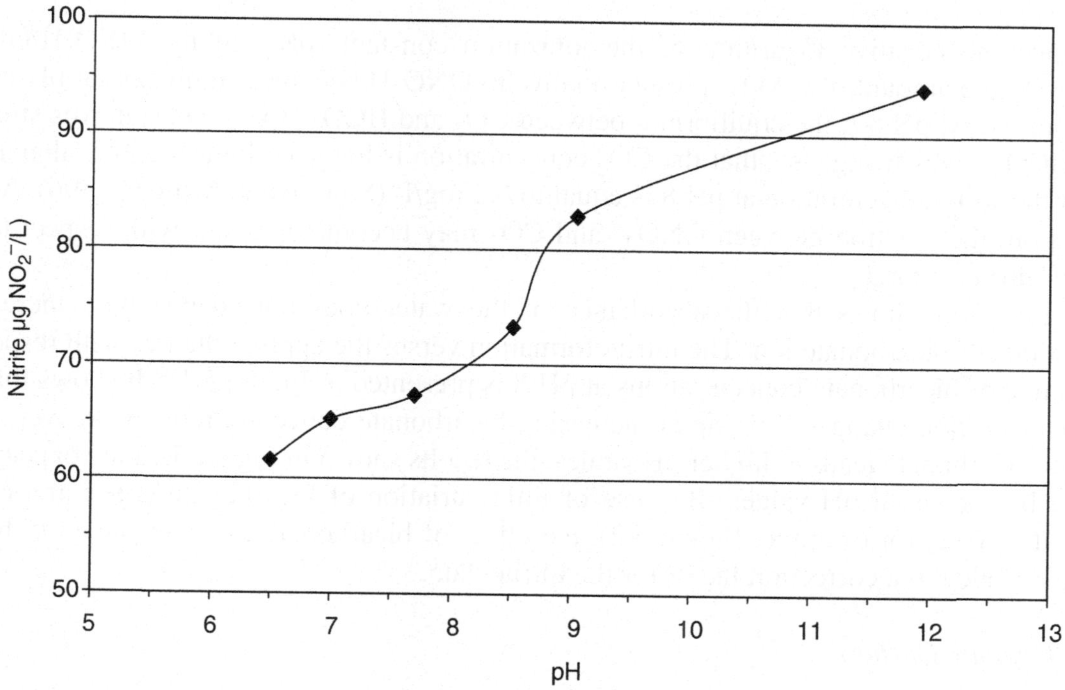

Figure 5.10 Nitrite formation versus pH at 400 J/m^2

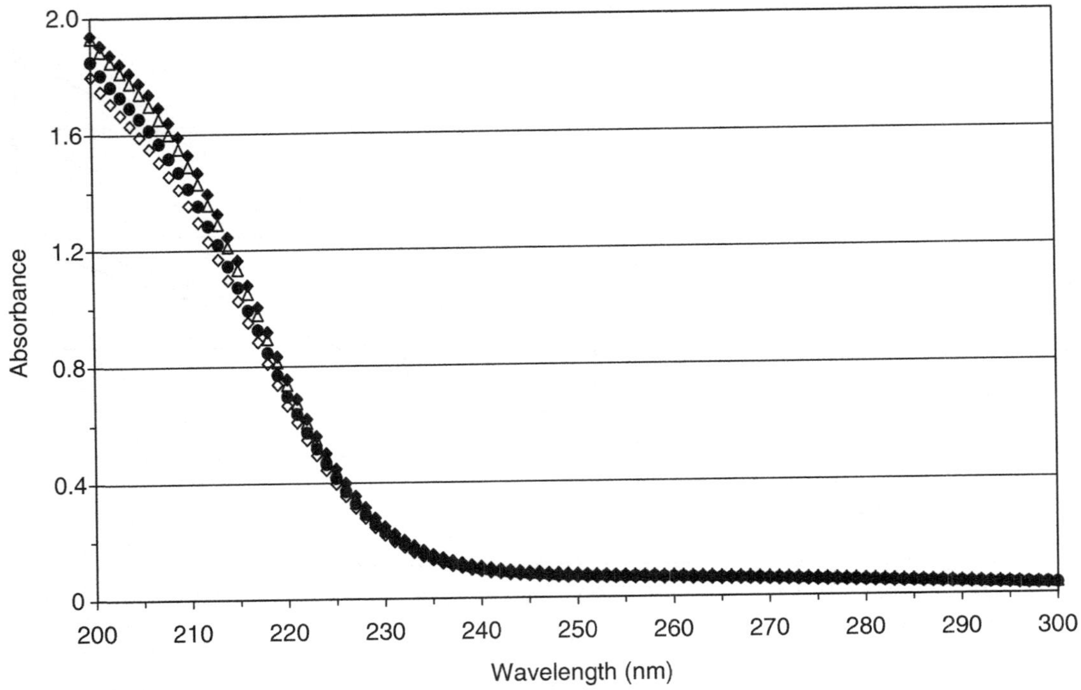

Figure 5.11 Change in absorbance of the pretreated IJssel Lake water with bicarbonate content (mg/L): 136 (♦), 171 (△), 204 (●), 238 (◇). Water quality: 11.2 mg NO_3^-/L, pH = 7.80–8.00, NPOC = 3.6 mg C/L.

Since the negative logarithm of the ionization constant (pK_a) of the ONO_2H/ONO_2^- system is 6.8, the amount of ONO_2^- present relative to ONO_2H becomes significant at pH values greater than ~7. At pH >7, the equilibrium between CO_2 and HCO_3^- ($pK_a = 6.4$) moves strongly towards HCO_3^-, which suggests that the CO_2 concentration is low (see Figure 3.3). Calculations show that the CO_2 concentration at pH 8 is equal to 0.2 mg/L (Stumm and Morgan 1996). At this concentration, the reaction between ONO_2^- and CO_2 may become apparent with respect to the amount of nitrite formed.

Figure 5.11 shows that the absorbance of the water does not change with increasing concentrations of bicarbonate ion. The nitrite formation versus the applied fluence with respect to a broad range of bicarbonate concentrations at pH 8 is presented in Figure 5.12. It shows that the nitrite concentration changes little upon increasing bicarbonate concentrations. Moreover, since addition of bicarbonate leads to higher pH values, the results shown in Figure 5.12 are expected to be caused by increased pH values. Because of little variation of bicarbonate concentrations in natural waters (see, for example, Figure 5.2), the effect of bicarbonate ion is expected to be too low to be included in a correction factor for the nitrite data.

Effect of Organic Carbon

The nitrite quantum yield increases in the presence of oxidizable organic compounds. According to Mark and co-workers (1996), this is because of the protection of the $NO_2^\bullet$ radical by

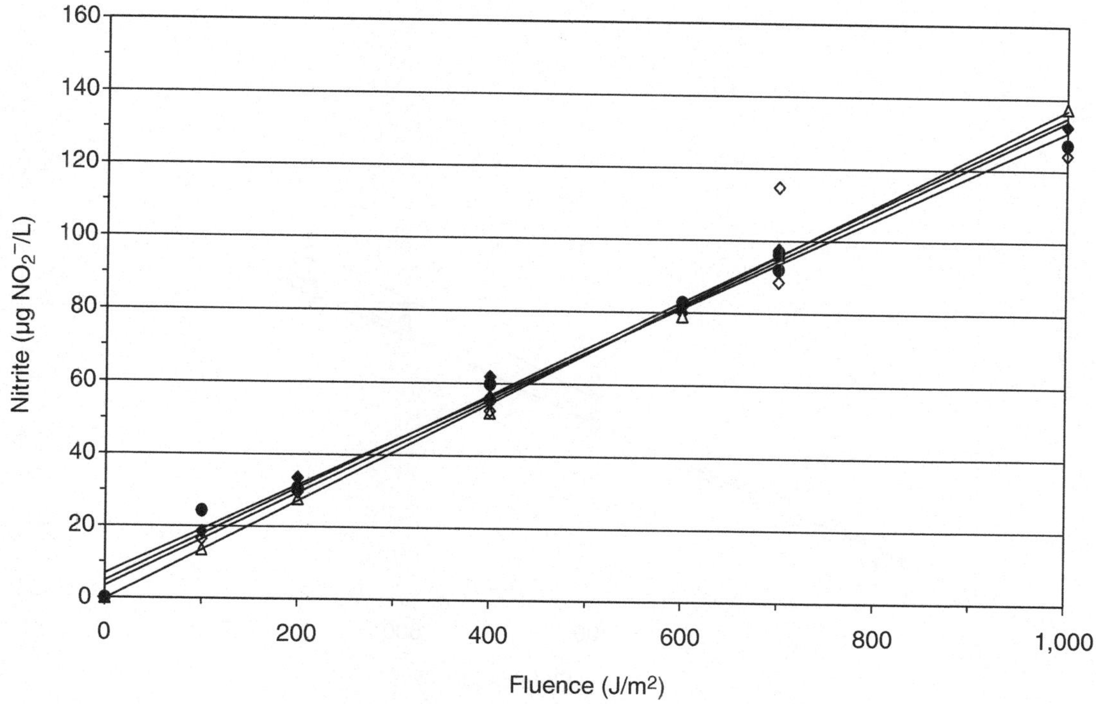

Figure 5.12 Nitrite formation versus the UV fluence for bicarbonate concentrations (mg/L) of 136 (♦), 171 (△), 204 (●), 238 (◇). Water quality: 11.2 mg NO_3^-/L, pH = 7.80–8.00, NPOC = 3.6 mg C/L.

the reaction of DOC with $O^{\bullet-}$ and $^{\bullet}OH$ (see reaction scheme in Figure 3.4). The change in absorbance versus DOC was already shown in Figure 5.4. Figure 5.13 shows the change in nitrite formation with increasing DOC concentration.

Although the absorbance decreases significantly in the 200–240-nm wavelength region upon addition of DOC, the formation of nitrite increases. This indicates that the (decrease in) UV transmission of the water is not responsible for the change in nitrite formation (Mark, et al. 1996; Sharpless and Linden 2001). The fact that the increase in nitrite formation is limited agrees with the relatively low amount of radicals formed at fluences for UV disinfection. Moreover, the DOC added to the water here ranged from 2 to about 13 mg C/L. This concentration range is much broader than the range in which DOC concentrations fluctuate over time in most natural waters. It is therefore expected that the small increase in nitrite formation with DOC content as shown in Figure 5.13 will be even smaller in natural waters. The situation may be different in cases of high nitrate concentrations and low DOC content.

Effect of Water Quality Variation

The absorbance and nitrite formation versus variation in water quality were monitored for a period of 1 year (see Figure 5.14 and Figure 5.15, respectively). Comparing the change in UV absorbance for different additions of nitrate (Figure 5.5) and in the pretreated IJssel Lake water over time (Figure 5.15), it can be seen that the decrease in UVT is roughly the same. This supports the expectation that changes in UV absorbance in this particular water are because of changes in

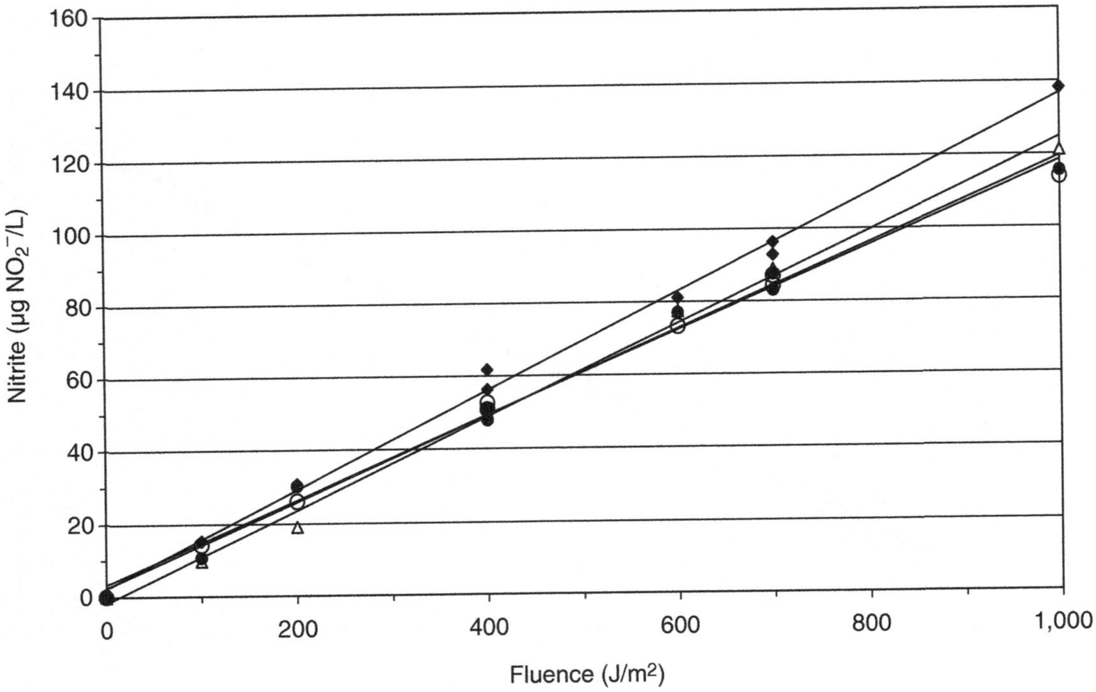

Figure 5.13 Nitrite formation versus the UV fluence for organic carbon (mg C/L): 2.1 (○), 3.1 (●), 6.2 (△), 13.4 (◆). Water quality: 10.3–10.9 mg NO_3^-/L, pH = 7.08–9.13, alkalinity = 205 mg HCO_3^-/L.

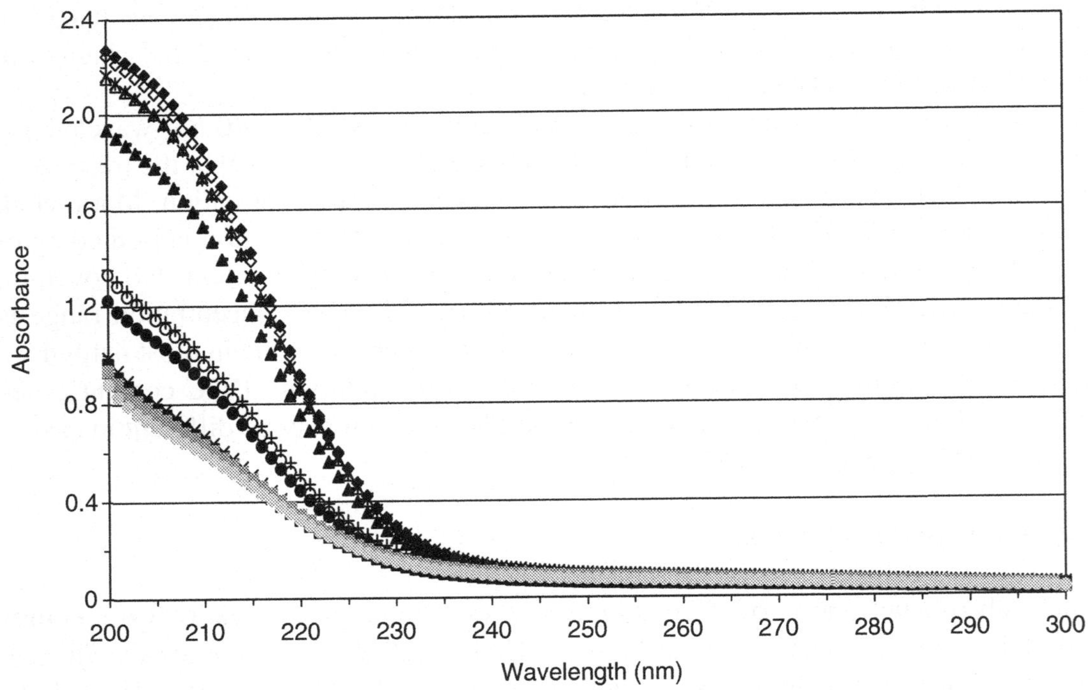

Figure 5.14 Change in absorbance of the pretreated IJssel Lake water over time. Legend based on nitrate concentration (mg NO_3^-/L): 3.5 (■), 3.5 (□), 3.9 (×), 5.7 (●), 6.6 (○), 11.2 (▲), 12.1 (△), 13.1 (◆), 12.3 (◇), 11.5 (⁎), 10 (–), 6.2 (+), 3.2 (▩), 2.9 (◎). Refer also to Table 4.1.

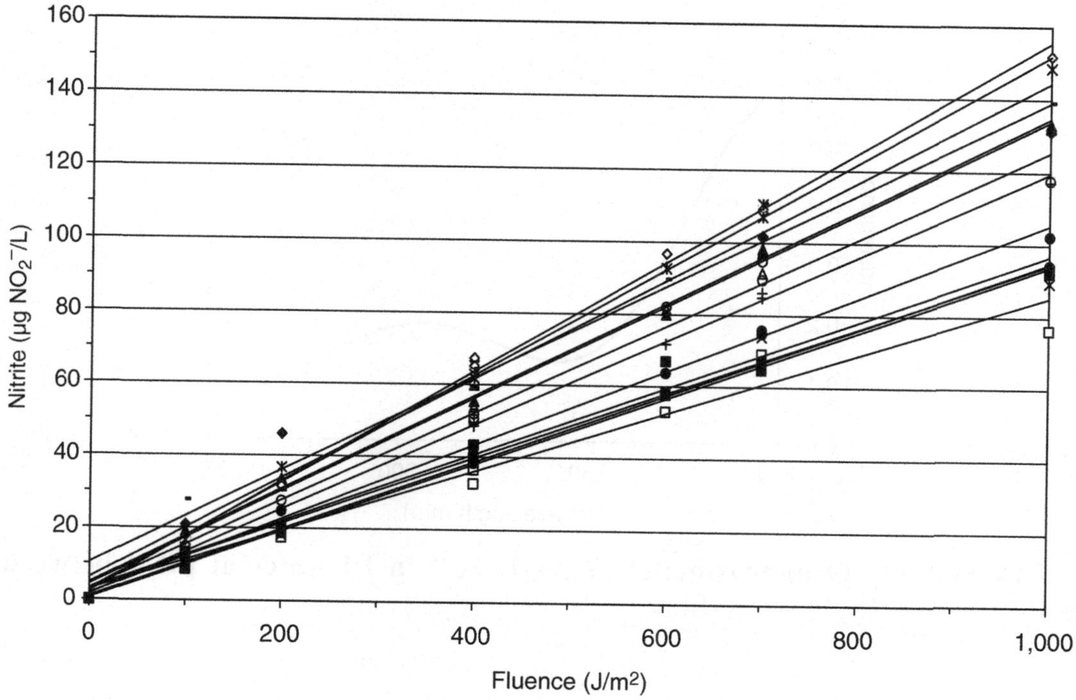

Figure 5.15 Nitrite formation versus the UV fluence over time with pretreated IJssel Lake water (see Table 4.2 for the water qualities). Legend based on nitrate concentration (mg NO_3^-/L): 3.5 (■), 3.5 (□), 3.9 (×), 5.7 (●), 6.6 (○), 11.2 (▲), 12.1 (△), 13.1 (◆), 12.3 (◇), 11.5 (∗), 10 (–), 6.2 (+), 3.2 (▦), 2.9 (◍).

nitrate concentration. However, it has to be elucidated whether this also counts for waters where the fluctuation in nitrate and DOC content is completely different from the water quality studied here (e.g., with fairly constant nitrate and strongly fluctuating DOC concentrations).

COMPOUNDS DOSED

The Fe^{3+}/Fe^{2+} System

Based on a previous literature report, it is expected that this system will be useful only at pH values below ~5 where Fe^{3+} is fairly stable in solution and the photo reduction rate is highest (Collienne 1983). In all the experiments performed here, the pH was set to 4. However, it is questionable whether such a pH is realistic for monitoring purposes in practice.

The initial Fe^{3+} concentration of 5 mg/L, used in most experiments, was chosen based on the model estimates of Fe^{2+} production rates. While this level of iron is well above the U.S. Environmental Protection Agency's (USEPA's) maximum contaminant level (MCL) of 0.5 mg/L, a brief literature search was unable to find enough information to judge whether it poses an acute toxicity risk. This topic is clearly of importance for further development of the Fe^{3+}/Fe^{2+} system as a fluence meter; and given the promising results shown here, it should be a priority to obtain more information on this subject.

One of the first concerns addressed before carrying out extensive work with this system was the stability of the Fe^{3+} in solution at pH 4. Theoretically, formation of oxide solids should be

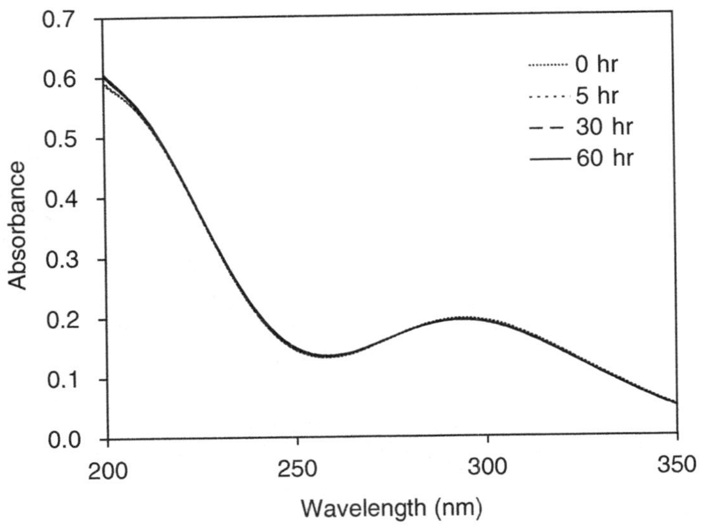

Figure 5.16 Stability of approximately 7 mg/L Fe^{3+} in DI water at pH 4 between 0 and 60 hours

negligible at this pH, and this was verified in DI water by monitoring the absorbance of the solution over 3 days (Figure 5.16).

The data in Figure 5.16 show that Fe^{3+} is dissolved throughout the course of all experiments presented here. Two points are worth noting. First, the stability of Fe^{3+} can also be expected in natural waters at pH 4 because of the influence of organic and inorganic ligands. Second, the addition of milligram-per-liter quantities of Fe^{3+} imparts a significant amount of UV absorbance to the water, which decreases the applied fluence compared with cases without iron; in addition, a slight coloration can be visually detected at 5 mg/L iron in DI water.

Figure 5.17 presents the results of UV experiments in both DI water and synthetic water for both LP and MP lamps with 5 mg/L dissolved iron. Under the conditions employed here, it appears that germicidal fluences as low as 100 and 50 J/m^2 can be accurately quantified for LP and MP, respectively.

Two important trends are present in Figure 5.17. First, the Fe^{2+} production rate is higher in synthetic water, which will be discussed further. Second, there is clear downward curvature in the plots, which could complicate the use of this system as a fluence meter. Such curvature could arise because Fe^{3+} is being consumed during the experiments. In fact, based on mass balance considerations, this must be the case. Significant amounts of Fe^{2+} are generated (tens of micromolars), amounting to approximately 10% to 25% of the original Fe^{3+} for LP and MP, respectively. Surprisingly, although this loss of Fe^{3+} is evident in absorbance spectra of irradiated DI water solutions, it is not that clear in absorbance spectra of irradiated synthetic water solutions (Figure 5.18).

Investigation revealed the possibility that curvature in the Fe^{2+} plots could be due to Fe^{2+} oxidation back to Fe^{3+}. The rate of Fe^{2+} oxidation by O_2 at pH 4 is slow relative to the time scale of these experiments, so any oxidation would be because of photogenerated oxidants. The fact that Fe^{2+} production is higher in synthetic than in DI water suggests that $^{\bullet}OH$ could be a potential oxidant that is being scavenged by the NOM. However, NOM has previously been observed to

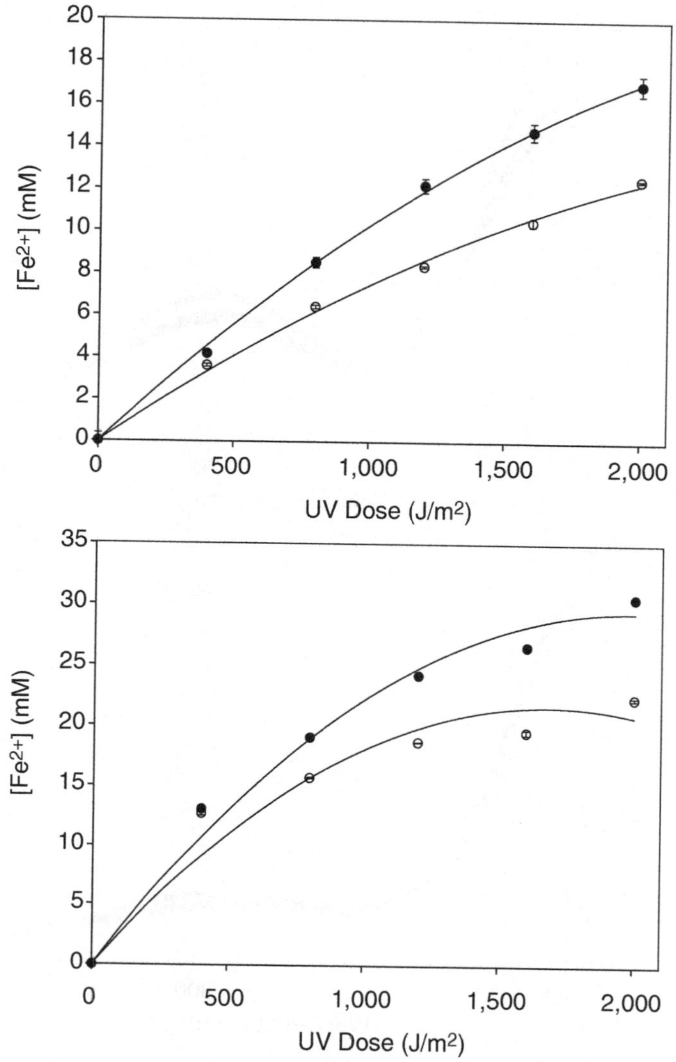

Figure 5.17 Fe^{2+} production in both DI and synthetic water with 5 mg/L Fe^{3+}: synthetic water (●), DI water (○). LP (top); MP (bottom).

accelerate Fe^{3+} reduction, presumably by direct e^- transfer in complexes or by production of $e^-(aq)$ (Collienne 1983).

In order to test whether $^{\bullet}OH$ could be an important oxidant, experiments were performed in DI water with the addition of varying quantities of 1-octanol, an excellent $^{\bullet}OH$ scavenger. If $^{\bullet}OH$ were oxidizing Fe^{2+}, the addition of the scavenger would eliminate this oxidation and reduce the apparent curvature. These data are presented in Figure 5.19. Surprisingly, the Fe^{2+} production rate decreases with octanol addition. This indicates that $^{\bullet}OH$ cannot be the primary oxidant of Fe^{2+}. These data are presently unexplained and present an opportunity for further study, which is beyond the scope of this project.

Because of the high concentrations of Fe^{2+} that were generated in the 5-mg/L Fe^{3+} solutions, experiments were also conducted in synthetic water to examine the feasibility of dosing 1 mg/L Fe^{3+}. The results of this experiment using the MP system are shown in Figure 5.20. Again, there is clear curvature due to the loss of approximately 25% of the original Fe^{3+}. Nonetheless,

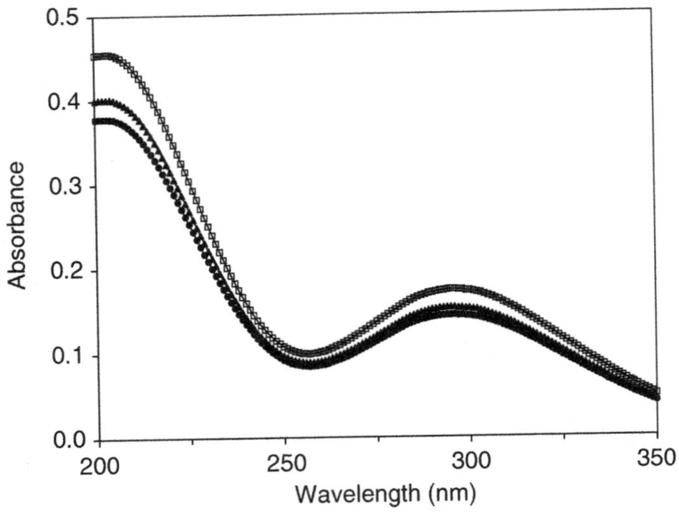

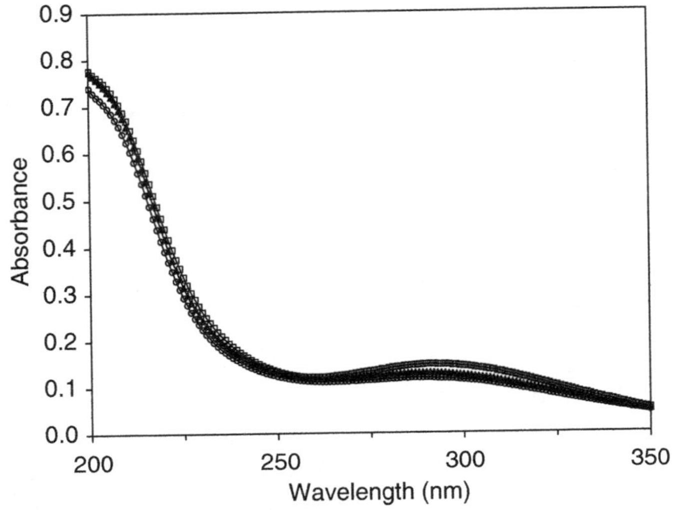

Figure 5.18 Change in absorbance with MP irradiation of 5 mg/L Fe^{3+} solution: 0 J/m^2 (□), 1,200 J/m^2 (▲), 2,000 J/m^2 (◆). DI water (top); synthetic water (bottom).

these data suggest that still lower concentrations of Fe^{3+} would be feasible for use as a fluence meter at low pH (4–5). Given an Fe^{2+} quantitation limit of 0.2 μM, the data in the figure suggest that substantially lower Fe^{3+} concentrations could be used to quantify polychromatic germicidal fluences, likely even below the 0.5-mg/L MCL set by USEPA. Experiments have not been conducted with the LP system at this lower Fe^{3+} concentration, but based on the data for 5 mg/L, it is anticipated that 1 mg/L is close to the lower Fe^{3+} limit that can be applied for monochromatic UV unless a cuvette with a longer path length were used or a more sensitive analytical method were found.

Fe^{3+} Experiments With Natural Waters

Experiments were conducted with the MP lamp using Durham water obtained in March, which had an intrinsic Fe^{3+} level of 0.5 mg/L (see chapter 4, Materials and Methods). These

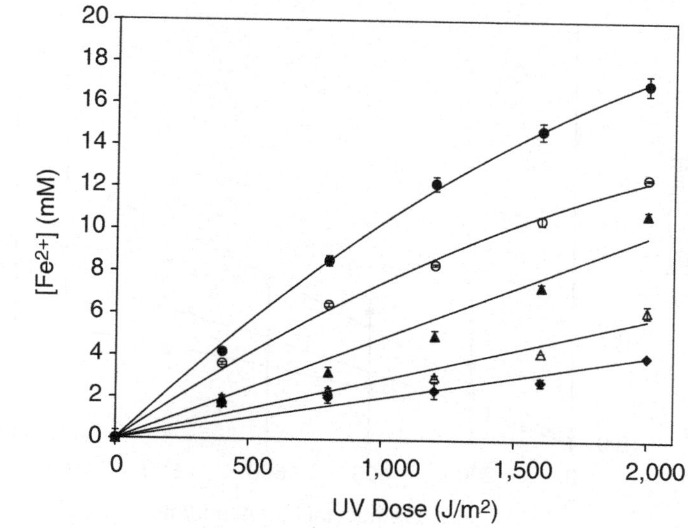

Figure 5.19 Effect of 1-octanol on Fe^{2+} production in buffer with LP lamp at 5 mg/L Fe^{3+} initial concentration: synthetic water (●), DI water (○), DI water with 5 μM octanol (▲), DI water with 10 μM octanol (△), and DI water with 20 μM octanol (◆)

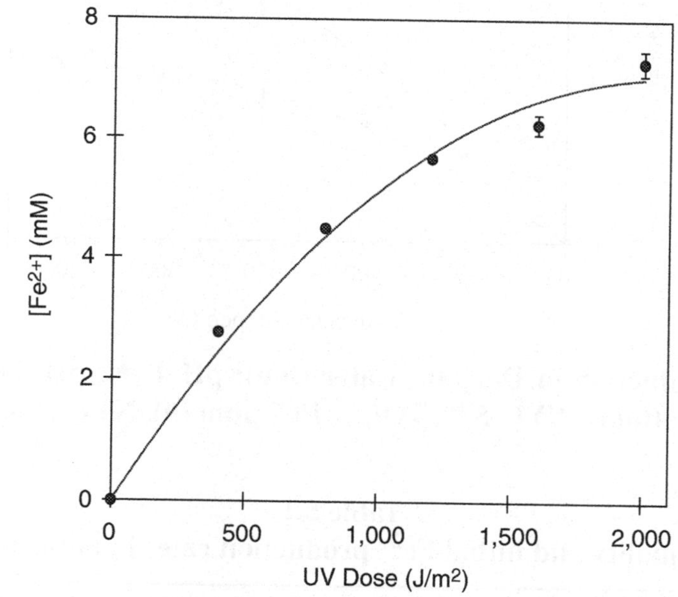

Figure 5.20 Fe^{2+} production in synthetic water with MP at 1 mg/L initial Fe^{3+} concentration

experiments were run at pH 4, 5, 6, and 7. Consistent with previous results using the synthetic water, no Fe^{2+} production was observed at pH 7. Measurable rates were obtained at pH 4, 5, and 6 (Figure 5.21), and rates decreased with pH as expected. Experiments were also conducted with the two NYC waters, which had different levels of TOC, alkalinity, and hardness (Table 5.1). These experiments involved spiking Fe^{3+} at 1.0 mg/L. The production of Fe^{2+} was greatest in the water with lower TOC, as shown in Figure 5.21 and Table 5.1.

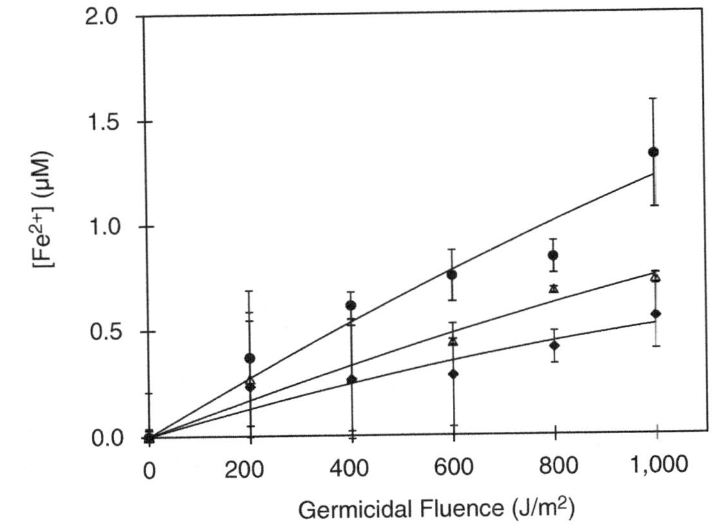

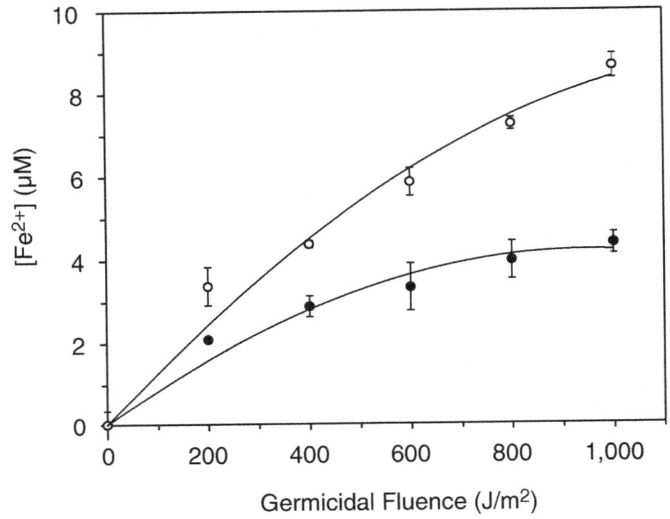

Figure 5.21 Fe^{2+} production in Durham water (top): pH 4 (●), pH 5 (△), pH 6 (♦); and in NYC waters (pH 6, bottom): NYC-579, TOC = 16.7 ppm (●), NYC-570, TOC = 3.2 ppm (○)

Table 5.1
Water quality and initial Fe^{2+} production rates in natural waters

	Initial Fe^{2+} production rate ($\mu M\ m^2\ J^{-1}$)	$[Fe^{3+}]^0$ (mg/L)	TOC (mg/L)	Alkalinity (as $CaCO_3$)	Hardness (as $CaCO_3$)	Nitrate (mg N/L)
Durham, pH 4	$1.4 \cdot 10^{-3}$	0.5*	9.1	17	44	0.67
Durham, pH 5	$8.8 \cdot 10^{-4}$	0.5*	9.1	17	44	0.67
Durham, pH 6	$6.9 \cdot 10^{-4}$	0.5*	9.1	17	44	0.67
NYC-579, pH 6	$8.9 \cdot 10^{-3}$	1.0†	16.7	55.5	175	0.53
NYC-570, pH 6	$1.3 \cdot 10^{-2}$	1.0†	3.2	7	16	0.47

* 0.5 mg/L Fe^{3+} present in Durham water without spiking.
† 1.0 mg/L Fe^{3+} spiked into NYC water.

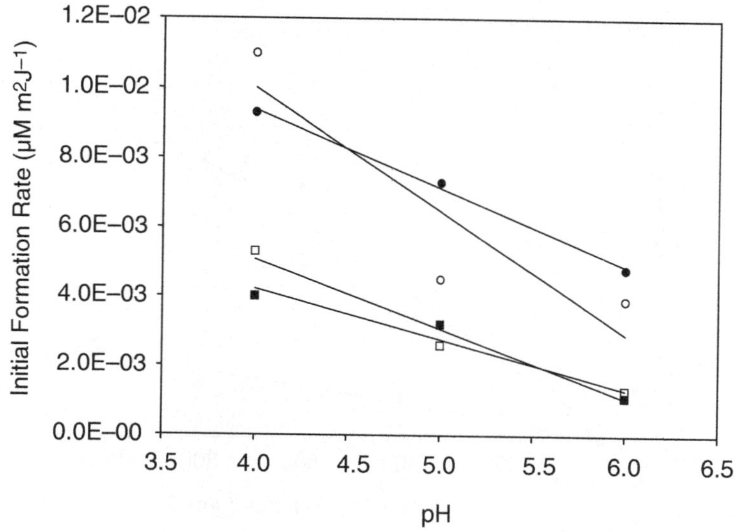

Figure 5.22 Initial rates of Fe^{2+} production in synthetic water: 0.5 mg/L immediately measured (▪), 0.5 mg/L after equilibrium (▫), 1 mg/L immediately (•), and 1 mg/L equilibrium (◦)

Based on previous results with synthetic water, the initial Fe^{2+} production rate in the Durham water was lower than expected. Figure 5.22 shows the results obtained with synthetic water (TOC = 3 mg/L). The pH dependence of the Fe^{2+} production rates was also weaker in the Durham water than in the synthetic water. These results may be due to complexation of the intrinsic Fe^{3+} by NOM in the Durham water, but no conclusions about this theory can be drawn from the present data.

In contrast to the Durham water, the Fe^{2+} production rates in NYC water were much higher than expected based on the synthetic water results (Figure 5.22 and Table 5.1). Again, the reasons for this are not presently known. The other major result from the NYC waters is the lower rate of Fe^{2+} production in the water with higher TOC. While this water also has higher levels of alkalinity and hardness, the trend with increasing TOC agrees with previous results using 1-octanol in DI water (Figure 5.19). The results with 1-octanol were obtained with 5 mg/L Fe^{3+}, whereas the NYC waters contained 1 mg/L Fe^{3+}. Nonetheless, the 1-octanol results suggest that the effect of TOC should decrease at TOC levels greater than 2 mg/L. This is qualitatively what is observed with the NYC waters, where, despite a very large difference in TOC levels, the Fe^{2+} production rates differ by only 44%.

The present results indicate substantial effects of water quality on Fe^{2+} production rates, and it seems unlikely that the present data set is sufficient to quantitatively evaluate these effects.

Similar experiments were conducted using Durham water obtained in June 2004 from the Williams Treatment Plant. This water did not contain detectable iron, in contrast to the water obtained in March. Therefore, 0.5 and 1 mg/L of Fe^{3+} were spiked for the experiments.

Kinetic Analysis for Natural Waters

To more accurately quantify Fe^{2+} production kinetics, a first-order reaction model was applied. This reaction order is based on first-order photolysis of Fe^{3+} and leads to the following expression for the Fe^{2+} concentration versus average fluence, E'_{avg}:

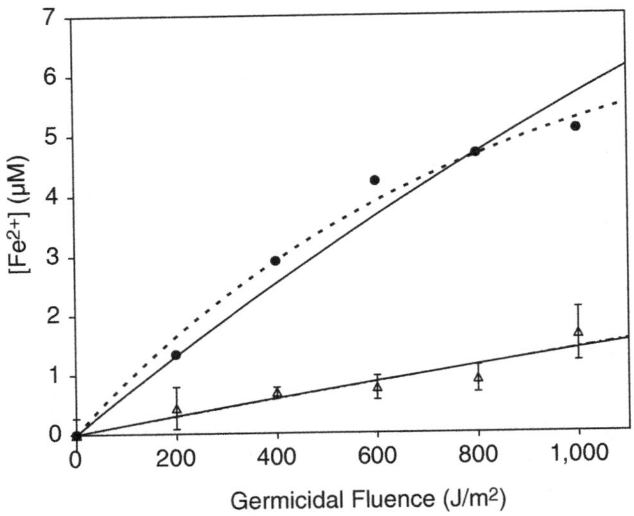

Figure 5.23 Nonlinear regression fits to data obtained for Durham water (June sample) at pH 5: 0.5 mg Fe^{3+}/L (△), 1 mg Fe^{3+}/L (•). The dashed lines are the fits to Equation 5.1, whereas the solid lines are fits of second polynomials such as used in previous experiments with DI and synthetic water.

$$[Fe^{2+}] = [Fe^{2+}]_{\infty}(1 - e^{-k'_f E'_{avg}}) \qquad (5.1)$$

Here, $[Fe^{2+}]_{\infty}$ indicates the concentration at infinite time (the steady-state concentration), and k'_f is a fluence-based rate constant for the reaction. This equation may be fit to the data with $[Fe^{2+}]_{\infty}$ set as either a fitting parameter or a fixed value estimated from the data. A fit of this type is shown in Figure 5.23.

The data gathered in these experiments did not cover a sufficient range of Fe^{2+} concentration to always obtain a reasonable fit of the equation with $[Fe^{2+}]_{\infty}$ as a fitted parameter. Thus, the data were fit assuming that this value is equal to the original, starting, value of $[Fe^{3+}]$ (i.e., it was assumed that at infinite time, all the iron is reduced). The rate constants derived from this analysis are given in Table 5.2 for experiments with synthetic water and both samples of Durham water (March and June).

The rate constants found in these experiments all lie in a similar range regardless of the water investigated. However, no clear relationship between water quality, or Fe^{3+} concentration, and rate constants can be seen. For example, the water quality for the two Durham water samples is not dramatically different (Table 4.2), but both the absolute values of the rate constants and their sensitivity to pH are clearly different. Also, although the rate constants in the June sample of Durham water vary approximately linearly with Fe^{3+} concentration, this is not the case for the synthetic water. Investigating these phenomena would be an interesting research problem, but further work was not pursued with this system for two reasons. First, the fact that the system can only be applied below pH 6 makes it unlikely to be a universally useful fluence meter in the field. Second, the complications involved in modeling the behavior of the system as a function of experimental variables were beyond the scope of this work.

Table 5.2
Rate constants derived from nonlinear regression to first-order model for Fe^{2+} production

	pH	k (m^2 J^{-1})	
		0.5 mg Fe^{3+}/L	1.0 mg Fe^{3+}/L
Synthetic water	4	$3.75 \cdot 10^{-4}$	$4.31 \cdot 10^{-4}$
	5	$3.36 \cdot 10^{-4}$	$1.91 \cdot 10^{-4}$
	6	$8.61 \cdot 10^{-5}$	$1.30 \cdot 10^{-4}$
Durham water (March 2004)	4	$1.48 \cdot 10^{-4}$	—
	5	$8.98 \cdot 10^{-5}$	—
	6	$6.27 \cdot 10^{-5}$	—
Durham water (June 2004)	4	$2.35 \cdot 10^{-4}$	$4.21 \cdot 10^{-4}$
	5	$1.72 \cdot 10^{-4}$	$3.81 \cdot 10^{-4}$
	6	$1.15 \cdot 10^{-4}$	$3.65 \cdot 10^{-4}$

The H_2O_2/Benzoate System

Experiments with a range of H_2O_2 concentrations at a fixed benzoate concentration of 50 μM were investigated with this system. This concentration of benzoate was chosen for two reasons. First, at the lower rates of 4-hydroxybenzoate formation predicted by modeling (and anticipated in natural waters), it was expected that the benzoate concentration would remain essentially unchanged at germicidal fluences up to approximately 1,000 J/m^2, which greatly simplifies the data analysis. Second, this level of benzoate is essentially nontoxic; data available on the World Health Organization's (WHO's) Web site indicates that adults may safely consume an upper limit of 5 mg kg^{-1} day^{-1} of benzoate (WHO 2000). At 50 μM, this corresponds to roughly 45 L of water for a 120-lb (55-kg) person. Sodium benzoate is a very common preservative used in many foods and beverages, but it is difficult to ascertain the exact concentrations without contacting specific manufacturers. Regarding the products' hydroxybenzoates, very little information could be located on their toxicity, but based on material safety data sheet (MSDS) information, they appear to be even less toxic than benzoate itself. This is clearly a topic for more literature research in the interest of developing this system further for use as a fluence meter.

Figure 5.24 presents the results of experiments carried out in phosphate buffer at pH 7. During these experiments, the H_2O_2 concentration remained constant within experimental error.

As in the experiments with Fe^{3+}, the plots display negative curvature, indicating either a decreasing benzoate concentration or destruction of 4-hydroxybenzoate or both. The absorbance spectra of the MP-irradiated solutions (Figure 5.25) indicate substantial loss of benzoate in solutions containing H_2O_2. This can also be shown by HPLC analysis (data not shown).

The data collected to date do not, however, rule out the possibility that 4-hydroxybenzoate may be reacting with $^{\bullet}OH$.

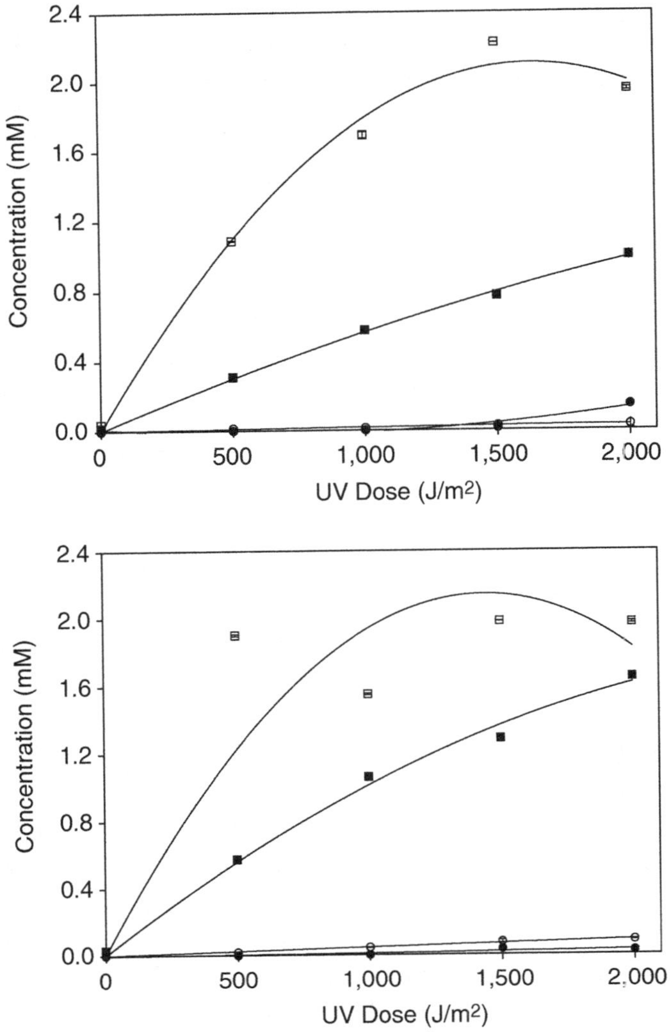

Figure 5.24 4-hydroxybenzoic acid production in phosphate buffer with LP (top) and MP (bottom). H_2O_2 concentration: 0 mg/L (●), 1 mg/L (○), 5 mg/L (■), and 10 mg/L (□).

Figure 5.26 presents the results of experiments performed in synthetic water at pH 8. The main difference between these results and those in buffer is a much lower rate and extent of 4-hydroxybenzoate formation. The same curvature is seen in the plots as in the experiments with buffer, but to a much lesser degree. In fact, the data appear to be roughly linear up to 1,000 J/m^2 under most conditions. Consistent with this is a much lower rate of benzoate loss as evidenced in the MP-irradiated solutions' absorbance spectra (Figure 5.27).

From the data in Figures 5.24 and 5.26, some preliminary conclusions can be drawn concerning the appropriate level of H_2O_2 to use. In buffer, 5 mg/L H_2O_2 is more than sufficient, but because the primary $^{\bullet}$OH scavenger in this solution is benzoate, it hardly represents a realistic treatment scenario. In synthetic water (TOC approximately 2.5 to 3 mg/L), appropriate H_2O_2 levels appear to be 5 mg/L for MP and 10 mg/L for LP. These concentrations generate sufficient product to quantify accurately, they are low enough under the experimental conditions to minimize loss of benzoate, and they lead to only slight deviations from linearity in the plots.

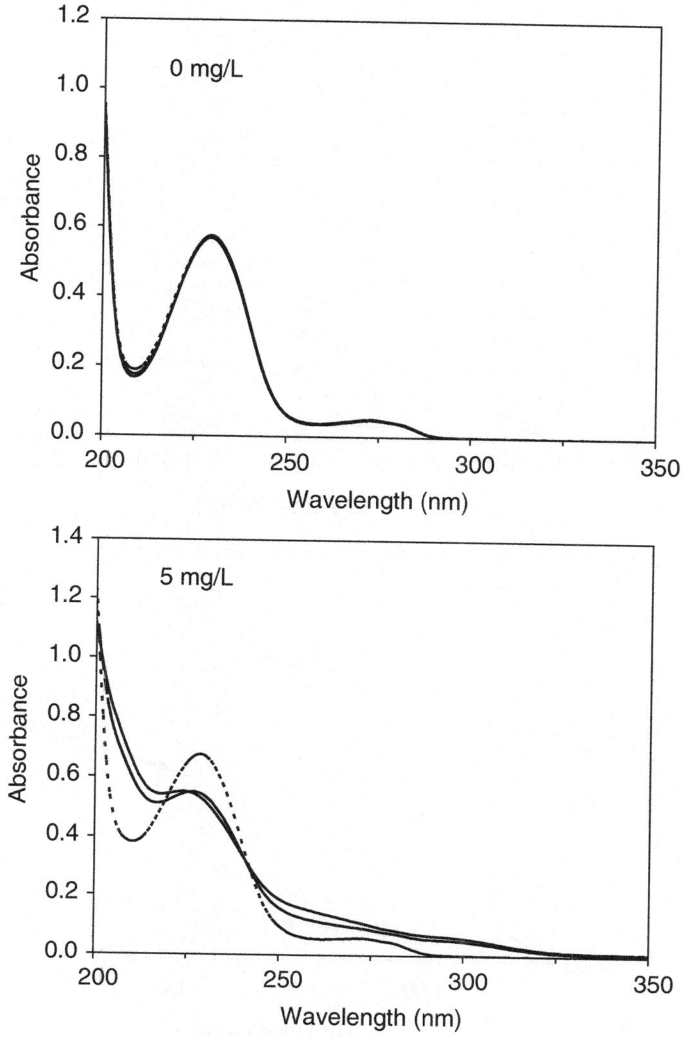

Figure 5.25 Stability of benzoic acid during MP irradiation in buffer with different H_2O_2 levels: 0 J/m^2 (----), 1,000 J/m^2 (– -), and 2,000 J/m^2 (—)

Some experiments with Durham water (March 2004 sample) were pursued, but the scavenging of $^{\bullet}OH$ by this water was so large that hydroxybenzoic acid could not be detected using any of the concentration regimes described above. At this point it was decided that pursuing further research along these lines was unwarranted. The basis for this decision was a comparison of the relative merits of the nitrate/nitrite and H_2O_2/benzoate systems (see chapter 7). Briefly, the data analysis and analytical methods required for the nitrate system are much less complicated and expensive, respectively. In addition, potential issues relating to the health and safety of adding H_2O_2 and benzoate to drinking water are difficult to foresee. In all, while this system may have useful applications in research, it is clearly not ideal as a fluence meter.

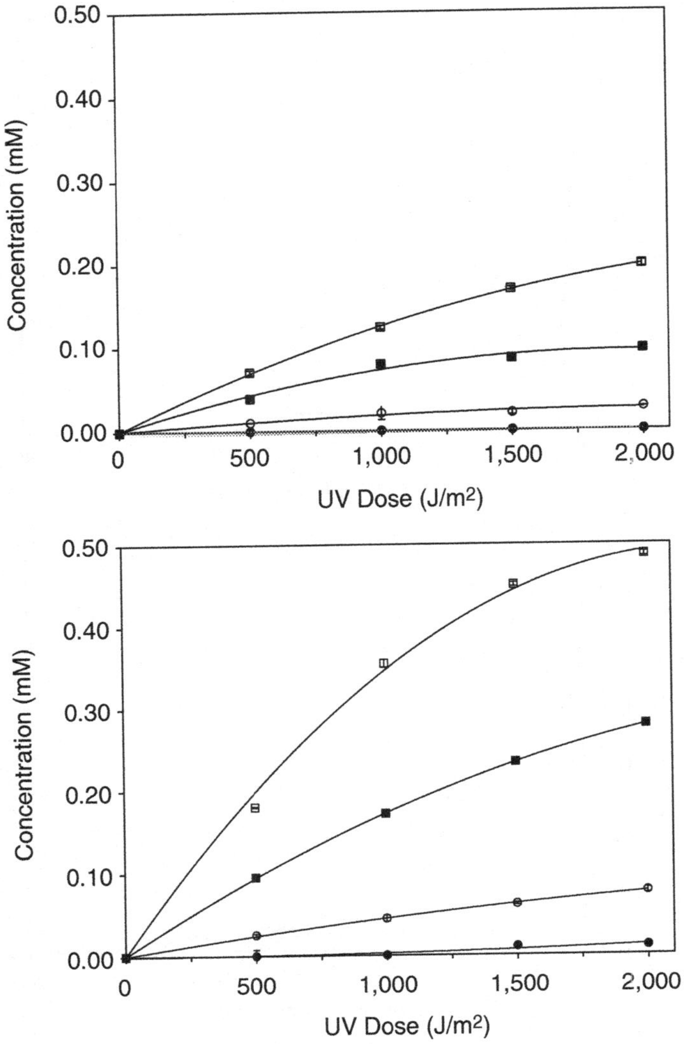

Figure 5.26 4-hydroxybenzoic acid production in synthetic water with LP (top) and MP (bottom). H_2O_2 concentration: 0 mg/L (●), 1 mg/L (○), 5 mg/L (■), and 10 mg/L (□).

SUMMARY

With respect to practical application of a fluence meter, it was concluded that nitrate–nitrite conversion is the best option. Compared with the Fe^{3+}/Fe^{2+} system and the H_2O_2/benzoate system, developing a fluence meter applying the nitrite formation level has certain advantages:

- Nitrate is available in most natural waters.
  - No addition of chemicals.
  - Compound is well mixed.
- Nitrite is the only photo product. The nitrite concentrations can be analyzed in measurable quantities that are well below the drinking water standard using UV fluences for disinfection (United States: 3.3 mg/L; European Union: 0.1 mg/L).

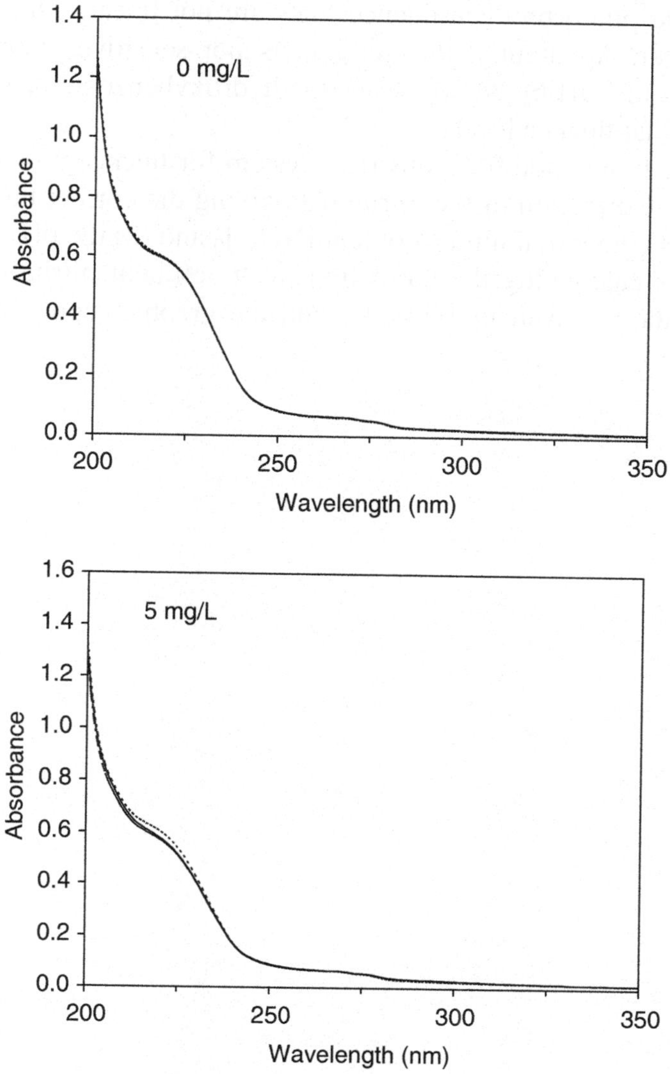

Figure 5.27 Stability of benzoic acid during MP irradiation in synthetic water with different H_2O_2 levels: 0 J/m^2 (----), 1,000 J/m^2 (– -), and 2,000 J/m^2 (—)

- Nitrite formation is linear with the fluence in the fluence range applied for UV disinfection (≤1,000 J/m^2).
- Nitrite levels as low as about 0.1 μg/L (as N) can be analyzed online.

The disadvantage of the nitrate–nitrite system is the significant effect of variation in water quality on nitrite formation. These effects must be quantified enabling a correction of the nitrite data with respect to the fluence delivered to the water.

Major drawbacks of the Fe^{3+}/Fe^{2+} system are the required low pH (6 or lower), the uncertainties surrounding water quality effects and the dependence of Fe^{2+} production rate on Fe^{3+} concentration, and the low concentration of iron present in most natural waters. The latter implies that iron may have to be dosed. The main disadvantages of the H_2O_2/benzoate system is the addition of H_2O_2 and benzoic acid and the advanced analysis methods required to measure the photo products. Moreover, the photo product–fluence lines are not linear, which makes it difficult to develop a fluence meter. Apparently, this method is not sensitive enough in natural waters, because the scavenging of $^{\bullet}OH$ by NOM causes the hydroxybenzoate production rates to be too low to quantify germicidal fluence levels.

In chapter 6 the foundation for a practical system for fluence monitoring based on nitrite formation is described. Correction of the nitrite data giving the correct fluence will be discussed using the variation in (1) the initial nitrate content $[NO_3^-]_0$ and (2) the pH.

Two procedures calculating the correction in the effluent nitrite data are proposed and exercised using the data set given in Table 4.1 and the graphic representation in Figures 5.14 and 5.15.

CHAPTER 6
PROPOSED SYSTEM FOR FLUENCE MONITORING

INTRODUCTION

It is clear from the previous chapter that a practical fluence meter system operating on the photolytic conversion of nitrate to nitrite may work, provided that the nitrite data are accurately corrected for variation in water quality. If the requirement of a minimal formation of nitrite for an accurate analysis is met, the actual nitrite formation level does not play a role.

It can be concluded that the nitrite data are more or less subject to variations in nitrate and DOC content, pH, and bicarbonate concentration. Variations in UV absorbance in the 200–240-nm range result from changes in nitrate and DOC content and will therefore be applied for correction of the variation in the nitrate and DOC concentration. The chemical effect of DOC on nitrite formation as found for the water studied in this research appeared to be small (see Figure 5.13). First, variation in bicarbonate ion may affect the nitrite formation by changing the pH of the water. Secondly, by acting as an $^{\bullet}OH$ scavenger, bicarbonate ion may change the nitrite level by decreasing the formation of $ONO_2^{\bullet}$ radical according to Reactions 3.9 and 3.10. However, the variation in bicarbonate ion concentration over time for the water studied here is relatively small, and OH radicals are only formed limitedly at the relatively low disinfection UV dose. The effect of radical formation and scavenging on nitrite formation was not further explored during this research and therefore will not be included in an initial proposed monitoring system. However, the effect of radical formation should be included in developing a nitrate fluence meter for UV/H_2O_2 applications.

The discussion above indicates that an initial practical monitoring system should be based on the variation in initial nitrate content ($[NO_3^-]_0$) and pH. This system is completely based on the results obtained with the pretreated IJssel Lake water quality studied in this research. Therefore, it must be established whether the mode of the monitoring system proposed here is applicable or acceptably accurate for other water qualities.

The previous chapter showed that the change in UV absorbance over time can be ascribed to the change in $[NO_3^-]_0$. Therefore, it is expected that monitoring the UV absorbance and calculating the surface area below the absorbance curves between 200 and 240 nm would correspond with the variation in $[NO_3^-]_0$.

By applying these relations, a procedure is proposed to quantify the effect of variation in water quality on nitrite formation. Depending on the applied procedure, a nitrite–fluence line is recorded to establish the initial settings. This procedure is proposed in the next section.

Monitoring $[NO_3^-]_0$, $[NO_2^-]$, pH, and UV absorbance (200–300 nm), the system builds its own data set. This data set and the next new input data will be used to correct the next nitrite data point.

PROPOSED MONITORING PROTOCOL

The following steps are proposed for monitoring:

1. Measure $[NO_2^-]$ in the influent and effluent water of the UV reactor giving $[NO_2^-]_{inf}$ and $[NO_2^-]_{eff}$, respectively.
2. Measure $[NO_3^-]_0$ in the influent water.

Table 6.1
Data set recorded with the natural water studied (see Table 4.1 and Figures 5.14 and 5.15)

Data point	Date	$[NO_3^-]_0$ (mg/L)	Surface area (a.u.)*	$[NO_2^-]_{meas}$ at 400 J/m^2 (µg/L)	pH
1	10-29-03	3.5	16.1	39.9	7.87
2	11-27-03	3.5	15.6	32.6	7.88
3	12-17-03	3.9	16.6	38.9	7.78
4	01-22-04	5.7	21.3	42.4	7.75
5	01-30-04	6.6	22.8	50.7	7.77
6	02-26-04	11.2	35.5	55.4	7.80
7	03-10-04	12.1	40.1	53.9	7.52
8	03-25-04	13.1	42.6	60.0	7.38
9	04-14-04	12.3	41.9	67.4	7.70
10	04-27-04	11.5	40.2	62.8	7.78
11	05-26-04	10.0	35.6	59.0	7.61
12	06-30-04	6.2	24.0	48.6	7.52

* Calculated from the absorbance spectra between 200–240 nm (see Figure 5.14). a.u. = arbitrary units.

3. Measure the pH of the influent or effluent water.
4. Correct $[NO_2^-]_{eff}$ for $[NO_2^-]_{inf}$ giving $[NO_2^-]_t$.
5. Scan the UV absorbance of the influent water between 200 and 300 nm.
6. Integrate the surface area (SA) below the UV abs. curve (200–240 nm) giving SA_{int}.
7. Establish the linear relation between SA and $[NO_2^-]_t$ at the average fluence delivered to the water, including an acceptable range (e.g., 95% reliability or 5% error).
8. Establish the linear relation between $[NO_3^-]_0$ and SA, including an acceptable range (e.g., 95% reliability or 5% error).

The monitoring protocol was applied using the main part of the data set collected with the IJssel Lake water studied in this research. One of the data points from this data set and the first data point outside this data set were considered an "old" and "new" data point, respectively, and used for the correction exercise in this chapter (see Table 6.1).

Figures 6.1 and 6.2 present the relation between the initial nitrate content ($[NO_3^-]_0$) and the SA below the UV-absorbance curves and the relation between SA and nitrite concentration ($[NO_2^-]_t$), respectively. It shows that the relation between $[NO_3^-]_0$ and SA is quite linear. Figure 6.3 shows the relation between the amount of nitrite formed and the pH. A nitrite–fluence line taken from the data set (covering the old data point) taken from Figure 5.15 is presented in Figure 6.4.

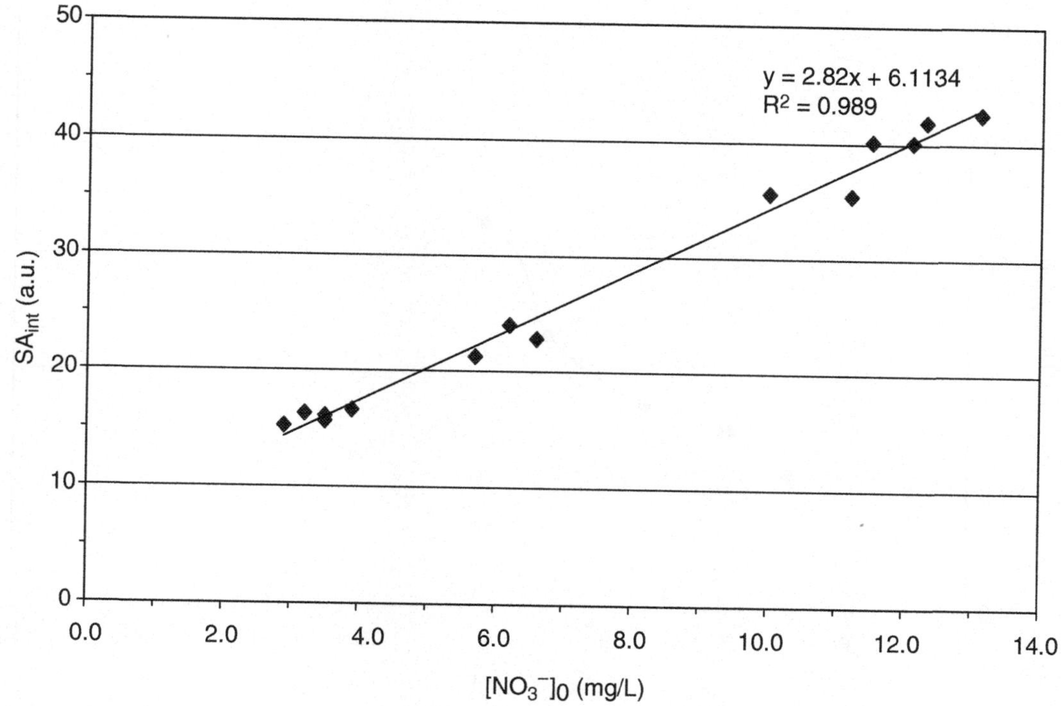

Figure 6.1 Surface area below the absorbance curves (200–240 nm; see Figure 5.14) versus the initial nitrate content $[NO_3^-]_0$ of the water

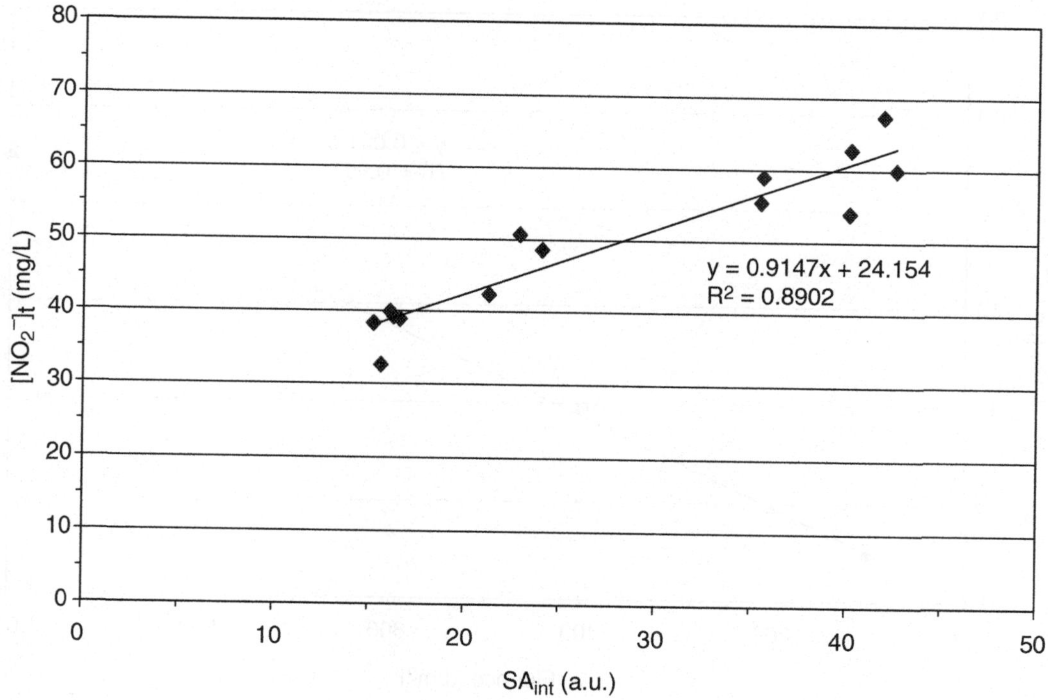

Figure 6.2 Nitrite formation at 400 J/m² versus the surface area (SA_{int}) below the absorbance curves (see Figure 5.14)

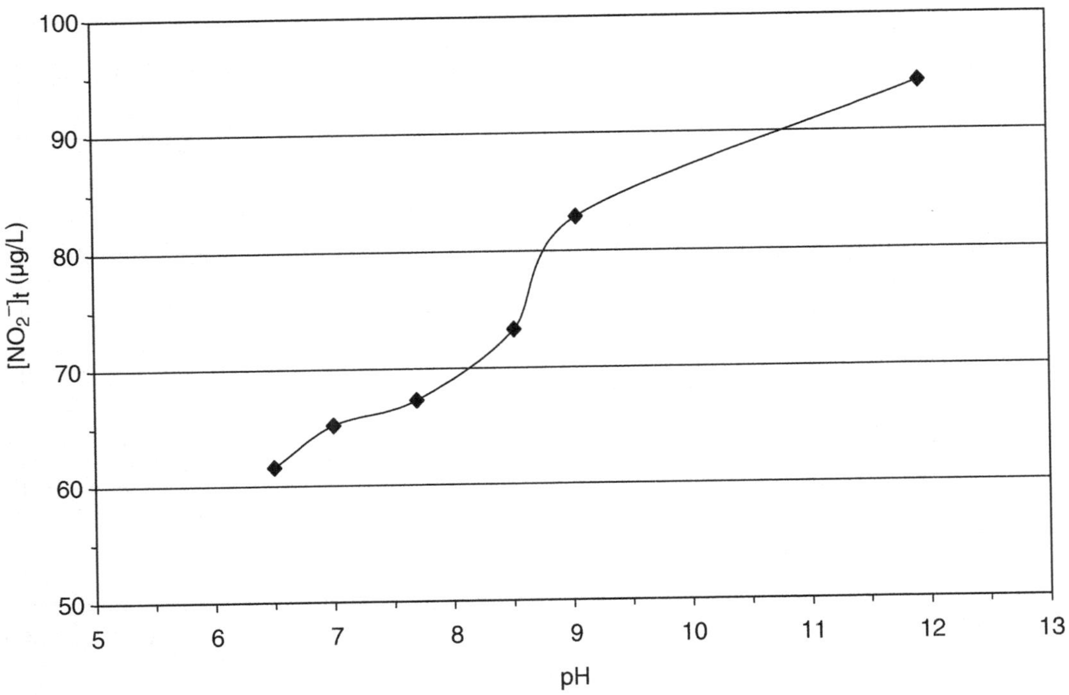

Figure 6.3 Relation between nitrite formation and pH to correct the nitrite concentration for pH changes at 400 J/m²

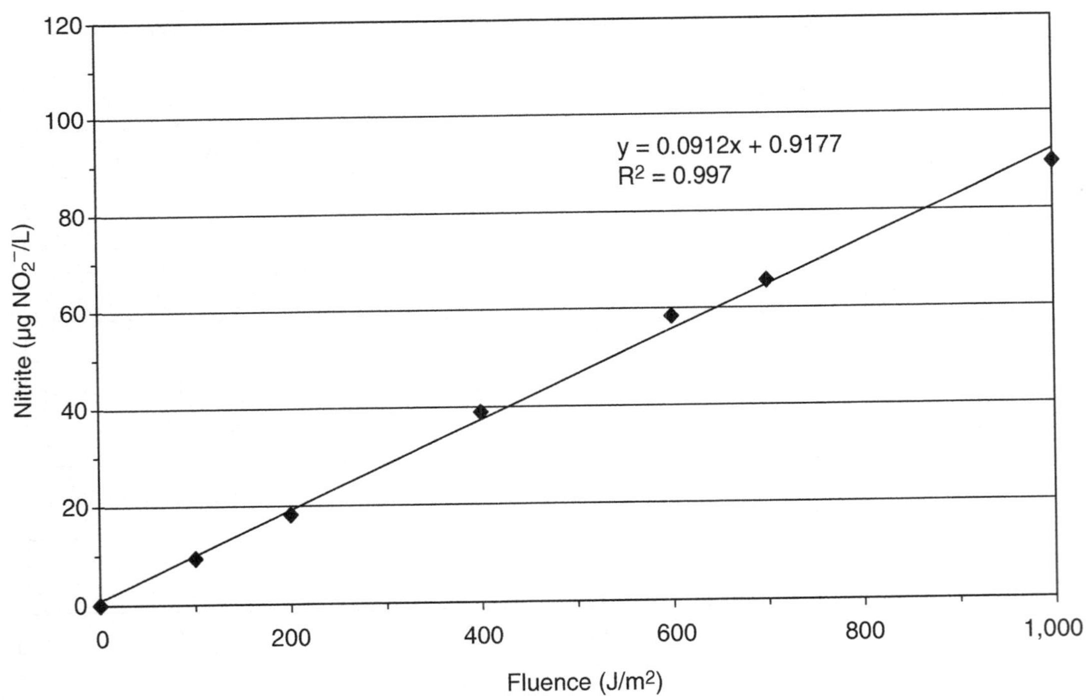

Figure 6.4 Nitrite–fluence line with $[NO_3^-]_0$ = 3.9 mg/L (see Figure 5.15 and Table 4.1)

DATA HANDLING

Collected data can be handled according to two procedures.

Calculation Procedure 1

The first procedure calculates the nitrite concentration using the collected data set and compares the corrected concentration ($[NO_2^-]_{corr}$) with the measured concentration ($[NO_2^-]_{meas}$). If the difference between those two numbers is within an acceptable range (e.g., 95% reliability), the fluence is acceptable.

1. Establish SA_{calc} from the Sa_{int} – $[NO_3^-]_0$ relation entering the new value for $[NO_3^-]_{0, new}$ (see Figure 6.1).
2. If SA_{calc} corresponds within an acceptable range with SA_{int}, then proceed with SA_{calc} (SA_{int} is included in the data set).
3. Establish $[NO_2^-]_{calc}$ from the $[NO_2^-]_t$–SA relation (see Figure 6.2).
4. Establish $[NO_2^-]_{pH}$ using pH_{new} and pH_{old}.
5. Correct $[NO_2^-]_{calc}$ for $[NO_2^-]_{pH}$ giving $[NO_2^-]_{corr}$.
6. Compare $[NO_2^-]_{corr}$ with $[NO_2^-]_{meas}$.
7. If $[NO_2^-]_{corr}$ is out of the range, an alarm will sound.

Calculation Procedure 2

An alternative procedure uses corrections between the last (subscript "old") and the new (subscript "new") nitrate and nitrite data points. The outcome is used to determine the fluence using a nitrite–fluence line obtained earlier with a collimated beam apparatus.

1. Establish SA_{new} and SA_{old} from the SA_{int}–$[NO_3^-]_0$ relation using the new $[NO_3^-]_{0, new}$ and the old $[NO_3^-]_{0, old}$, respectively (see Figure 6.1).
2. Establish $[NO_2^-]_{calc}$ entering SA_{new} in the $[NO_2^-]_t$–SA relation (see Figure 6.2).
3. Use SA_{new} and SA_{old} to determine $\Delta[NO_2^-]_{SA}$ from the $[NO_2^-]_t$–SA relation.
4. Establish $[NO_2^-]_{pH}$ using pH_{new} and pH_{old}.
5. Correct $[NO_2^-]_{calc}$ for $\Delta[NO_2^-]_{SA}$ and $[NO_2^-]_{pH}$ giving $[NO_2^-]_{corr}$.
6. Use $[NO_2^-]_{corr}$ to determine the fluence from the nitrite–fluence line (see Figure 6.4).

Calculation Exercise for Correction of Nitrite Data

A procedure was performed using the nitrate and nitrite concentrations and SA data that were measured over a period of 1 year (see Table 4.1 and Figure 5.15). Table 6.1 gives an overview of the data collected over time (in accordance with the monitoring protocol). The results of the corrections performed are shown in Table 6.2 (from calculation procedure 1) and Table 6.3 (from calculation procedure 2).

The results of redefining the fluence in Tables 6.2 and 6.3 indicate that the error is less than 5%. Calculating the fluence using the nitrite–fluence line shows a fluence of 435 J/m^2 for the new data point (12) which deviates about 5% from the 416 J/m^2 of the old data point (3).

Table 6.2
Results of the correction according to calculation procedure 1

Parameter	Units	Data point 3	Data point 12
$[NO_3^-]_0$	mg/L	—*	6.2
SA	a.u.	—	24.0
$[NO_2^-]_{calc}$	μg/L	—	46.1
pH	—	7.78	7.52
$[NO_2^-]_{pH}$	μg/L	—	1.3
$[NO_2^-]_{corr}$	μg/L	—	47.4
$[NO_2^-]_{meas}$	μg/L	—	48.6
Error	%	—	2.5

* Not applicable.

Table 6.3
Results of the correction according to calculation procedure 2

Parameter	Units	Data point 3	Data point 12
$[NO_3^-]_x$	mg/L	3.9	6.2
SA_x*	a.u.	16.6	24.0
$[NO_2^-]_{calc}$	μg/L	—†	46.1
$\Delta[NO_2^-]_{SA}$	—	—	−6.8
$[NO_2^-]_{pH}$	μg/L	—	+1.3
$[NO_2^-]_{corr}$	μg/L	—	40.6
$[NO_2^-]_{meas}$	μg/L	38.9	—
Fluence	J/m^2	416	435

* The subscript x represents the old data point (3) or the new data point (12).
† Not applicable.

CHAPTER 7
CONCLUSIONS

NATURALLY OCCURRING COMPOUNDS

Nitrate–Nitrite Conversion

Overview

Nitrate has the potential to be developed as a fluence meter. This results from the effective formation of nitrite at nitrate levels between 3 and 14 mg/L as studied using IJssel Lake water pretreated by coagulation, sedimentation, and rapid sand filtration. In addition, a linear relationship between the nitrite concentration after irradiation and the fluence is obtained at all possible IJssel Lake water qualities. Finally, nitrate is present in most natural waters, nitrite can be easily analyzed, and the nitrite analysis is inexpensive.

The formation of nitrite by irradiation of nitrate varies with the water quality. Therefore, a monitoring system is proposed to measure nitrate (influent), nitrite (influent and effluent), UV absorbance (influent), and pH (influent). Online techniques are available for these parameters, which makes the nitrate–nitrite conversion applicable for online fluence monitoring.

The nitrate–nitrite conversion takes place in the 200–240-nm range, which differs significantly from the 240–280-nm disinfection range. This may result in changes in the "nitrite fluence," whereas the disinfection or germicidal fluence (REF) is unchanged or vice versa. This may lead to false measures such as increasing the lamp power in order to correct the fluence. Therefore, it is expected that the nitrate fluence meter can be used as a warning system for UV disinfection systems. However, since OH radicals are produced during UV irradiation of hydrogen peroxide (H_2O_2) using 200–240-nm UV light, the nitrate fluence meter may be developed towards a full monitoring system, thereby enabling end users to control the UV fluence in UV/H_2O_2 systems.

Factors Affecting Nitrite Formation

The chemistry of nitrate photolysis is complex. The results from literature review, which have been largely confirmed in this collimated beam research, show that the nitrate content of the water and the pH play a significant role in the formation of nitrite. DOC and bicarbonate ion are known to affect the nitrite formation as well. By enabling fluence monitoring to the extent that nitrite is formed, the nitrite concentrations measured in the reactor effluent must be corrected for variations in the water quality.

As a result of agricultural activities, nitrate—as the primary source of nitrite formation during UV irradiation—may vary significantly over time in natural water. Therefore, the initial nitrate concentration $[NO_3^-]_0$ may be considered the most important parameter in nitrite formation. Affecting the equilibriums between the peroxynitrite anion (ONO_2^-) and its conjugated acid (ONO_2H) and between carbon dioxide (CO_2) and bicarbonate ion (HCO_3^-), the pH must be monitored as well. A lower pH shifts the chemical equilibrium toward ONO_2H and CO_2, thereby realizing a decrease in nitrite formation.

The effect of DOC increasing the nitrite formation is explained by the reaction of the organic molecules with radicals protecting the nitrite precursor. However, the increase in nitrite formation with increasing DOC concentrations was limited in the applied fluence range 0–1,000 J/m^2. Moreover, the range of DOC concentrations studied was much greater than that found in natural waters. Therefore, DOC is not included in the correction of the nitrite data in the proposed fluence monitoring system (see chapter 6).

Proposed Monitoring System

This research focused on the relative contribution of the individual parameters to nitrite formation in order to establish how to correct the nitrite data for variation in water quality. Given the strong effect of the initial nitrate content and, to a lesser extent, the pH on nitrite formation, a monitoring system based on these parameters is proposed.

Apparently, a linear relation exists between the area below the absorbance curve (200–240 nm) and the initial nitrate content of the water. Although the nitrate concentrations may vary between about 3 and 13 mg/L, this relation exists at all nitrate levels despite the increased relative contribution of DOC at the low nitrate levels. In addition to using the relation between nitrite formation and pH, the nitrite concentration in the effluent of the reactor is corrected for a shift in the chemical equilibriums.

However, if the DOC concentration varies to a large extent over time for a particular water source, the contribution of DOC to the absorbance curve may become apparent. This may result in a nonlinear relation between the area below the absorbance curve and the initial nitrate content, in which case the current proposed monitoring system would be unsatisfactory for this water quality.

The collimated beam data with respect to nitrate, nitrite (before and after UV irradiation), UV absorbance (200–240 nm), and pH of the pretreated surface water over a period of 1 year were applied to correct the nitrite data after UV irradiation for changes in these water quality parameters. Using these parameters, it was calculated that the change in nitrite concentration induced by a change in water quality could be corrected with a deviation of less than 5%. Since the result of this calculation procedure was obtained using the collimated beam data, verification of the procedure with a pilot UV unit is required.

COMPOUNDS DOSED

The Fe^{3+}/Fe^{2+} and the H_2O_2/Benzoate Systems

The Fe^{3+}/Fe^{2+} and the H_2O_2/benzoate systems are capable of quantifying fluence levels relevant to microbial pathogen inactivation. The results obtained in synthetic and natural waters indicate that the Fe^{3+}/Fe^{2+} system has the required sensitivity, but this is not the case for H_2O_2/benzoate in natural waters. Both systems suffer the slightly undesirable shortcoming where some degree of nonlinearity exists in the relationship between product and fluence. However, it appears that concentrations and fluence ranges can be judiciously chosen to minimize the difficulties posed by this. Furthermore, other data treatment schemes (e.g., first-order kinetic analysis) can be used to analyze the data.

The main differences between the systems concern the working pH range and the difficulty of the analytical method. For the Fe^{3+}/Fe^{2+} system, the pH must be at or below 6, whereas there is, in principle, no restriction on the pH range that can be used with H_2O_2/benzoate. In terms of analysis, the colorimetric Fe^{2+} analysis is much less costly and requires much less expertise on the part of the analyst than the hydroxybenzoate analysis. It is significant that several sensor designs have been published in the literature that incorporate sampling, mixing of sample and reagents, and spectroscopic analysis. These types of designs could, in principle, be applied to Fe^{2+} analysis online. Because the hydroxybenzoate analysis incorporates HPLC, online detection is not simple to achieve.

Another concern is the compliance of the parent and photoproduct compounds with the drinking water standards. For the Fe^{3+}/Fe^{2+} system, it appears that low concentrations of Fe^{3+} can be used ($\leq$1 mg/L) to eliminate any concerns, especially with the knowledge that the levels will only be this high during tests and that dilution will occur in storage and distribution. Furthermore, at typical pH values maintained in storage and distribution (>6), Fe^{2+}, which is relatively nontoxic, will be rapidly oxidized, whereby the only species of concern will be Fe^{3+}. For the H_2O_2/benzoate system, the necessary concentrations of benzoate are apparently low enough to eliminate toxicity concerns, but this should be investigated further. Also, the toxicity of the hydroxybenzoates should be clarified before further developing this system. Considering all of these issues, it becomes apparent that the nitrate/nitrite system has much better potential for cost-effective application to a wide range of waters than either the Fe^{3+} /Fe^{2+} or the H_2O_2/benzoate system.

REFERENCES

APHA, AWWA, and WEF (American Public Health Association, American Water Works Association, and Water Environment Federation). 1998. *Standard Methods for the Examination of Water and Wastewater*. 20th ed. Edited by L.S. Clesceri, A.E. Greenberg, and A.D. Eaton. Washington, D.C.: APHA.

Baxendale, J.H., and J.A. Wilson. 1957. *Transactions of the Faraday Society*, 53:344–356.

Bolton, J.R., and K.G. Linden. 2003. Standardization of Methods for Fluence (UV Dose) Determination in Bench-scale UV Experiments. *Jour. Environment Engineering*, (March): 209–215.

Bolton, J.R., and M.I. Stefan. 2002. Fundamental Photochemical Approach to the Concepts of Fluence Rate and Electrical Energy Efficiency in Photochemical Degradation Reactions. *Proc. of the IOA International Conference on Ozone in Global Water Sanitation*. Poitiers, France: IOA.

Chen, R. 1972. Measurements of Absolute Values in Biochemical Fluorescence Spectroscopy. *Jour. Research National Bureau Standards*, 76A(6):593–606.

Christman, R. 2002. Current Research in NOM Chemistry. Drinking Water Symposium, School of Public Health, University of North Carolina, Chapel Hill, N.C., May 22.

Collienne, R.H. 1983. *Limnology and Oceanography*, 28(1):83–100.

Du, H., A. Fuh, J. Li, A. Corkan, and J.S. Lindsey. 1998. PhotochemCAD: A Computer-Aided Design and Research Tool in Photochemistry. *Photochemistry and Photobiology*, 68:141–142.

Faust, B.C. 1994. *Aquatic and Surface Photochemistry*. Edited by G.R. Helz, R.G. Zepp, and D.G. Crosby. Boca Raton, Fla.: Lewis Publishers.

Gilmartin, M.A.T., J.P. Hart, and B.J. Bircg. 1994. Development of Amperometric Sensors for Uric Acid Based on Chemically Modified Carbon Epoxy-Resin and Screen-Printed Electrodes Containing Cobalt Phthalocyanine. *The Analyst*, 119:243.

Goldstein, S., and G. Czapski. 1998. Formation of Peroxynitrite From the Reaction of Peroxynitrite With CO_2: Evidence for Carbonate Radical Production. *Jour. American Chemical Society*, 120:3458–3463.

IJpelaar, G.F., and J.C. Kruithof. 2003. *Fluence Monitoring in UV Reactors—Desk Study of Chemical Substances to Be Developed as a Fluence Meter*. Kiwa report BTO 2003.054. Nieuwegein, the Netherlands: Kiwa Water Research.

Jagger, J. 1967. *Introduction to Research in Ultraviolet Photobiology*. Englewood Cliffs, N.J.: Prentice-Hall.

Kiwa Water Research. 2002. UV in Drinking Water Treatment—Scientific Foundation of Successful Application [CD-ROM]. UV Workshop, March 20–23. Nieuwegein, the Netherlands: Kiwa Water Research.

Klassen, N.V., D. Marchington, and H.C.E. McGowan. 1994. *Analytical Chemistry*, 66(18):2921–2925.

Lymar, S.V., and J.K. Hurst. 1995. Rapid Reaction Between Peroxynitrite Ion and Carbon Dioxide: Implications for Biological Activity. *Jour. American Chemical Society*, 117:8867–8868.

Lymar, S.V., and J.K. Hurst. 1996. Carbon Dioxide: Physiological Catalyst for Peroxynitrite-Mediated Cellular Damage or Cellular Protectant? *Chemical Research in Toxicology*, 9:845–850.

Mark, G., H-G. Korth, H-P. Schuchmann, and C. von Sonntag. 1996. The Photochemistry of Aqueous Nitrate Ion Revisited. *Jour. Photochemistry and Photobiology A*, 101:89–103.

Masschelein, W.J. 2002. *Ultraviolet Light in Water and Wastewater Sanitation*. Boca Raton, Fla.: Lewis Publishers.

Mathews, C.K., and K.E. van Holde. 1996. *Biochemistry*. 2nd ed. Menlo Park, Calif: Benjamin/Cummings Publishing Co.

Meulemans, C.C.E. 1987. The Basic Principles of UV-Disinfection of Water. *Ozone Science and Engineering*, 9: 299–314.

Minear, R.A. 1972. *Nucleic Acids in the Environment*. Edited by J.T. Trevors and J.D. van Elsas. New York: Springer-Verlag.

NEN (National Standards Institute). 1981a. NEN 6411. *Water: Determination of pH*. The Netherlands: NEN.

NEN (National Standards Institute). 1981b. NEN 6474. *Water Quality: Determination of Nitrite—Molecular Absorption Spectrometric Method*. The Netherlands: NEN.

Sharpless, C.M., K.D. Cho, and K.G. Linden. 2003. Investigation of Chemicals for Use as an On-line Fluencemeter for UV Reactors. AwwaRF Project 2949. Project Report 1. Denver, Colo.: AwwaRF.

Sharpless, C.M., and K.G. Linden. 2001. UV Photolysis of Nitrate: Effects of Natural Organic Matter and Dissolved Inorganic Carbon and Implications for UV Water Disinfection. *Environmental Science and Technology*, 35:2949–2955.

Sharpless, C.M., and K.G. Linden. 2003. Experimental and Model Comparisons of Low- and Medium-Pressure Hg Lamps for the Direct and H_2O_2 Assisted UV Photodegradation of *N*-Nitrosodimethylamine in Simulated Drinking Water. *Environmental Science and Technology*, 37:1933–1940.

Sharpless, C.M., M.A. Page, and K.G. Linden. 2003. Impact of Hydrogen Peroxide on Nitrite Formation During UV Disinfection. *Water Research*, 37:4730–4736.

Smalla, K., and J.D. van Elsas. 1995. *Nucleic Acids in the Environment*. Edited by J.T. Trevors and J.D. van Elsas. New York: Springer-Verlag.

Stryer, L. 1988. *Biochemistry*. 3rd ed. New York: W.H. Freeman and Company.

Stumm, W., and J.J. Morgan. 1996. *Aquatic Chemistry: Chemical Equilibria and Rates in Natural Waters*. 3rd ed. New York: Wiley-Interscience.

Tóth, M., Z. Kukor, and S. Valent. 2002. Chemical Stabilization of Tetrahydrobiopterin by L-Ascorbic Acid: Contribution to Placental Endothelial Nitric Oxide Synthase Activity. *Molecular Human Reproduction*, 8(3):271–280.

USEPA (U.S. Environmental Protection Agency). 1991. Method 300.1. *Determination of Inorganic Anions in Drinking Water by Ion Chromatography*. National Exposure Research Laboratory, Office of Research and Development. Cincinnati, Ohio: USEPA.

WHO (World Health Organization). 2000. *Benzoic Acid and Sodium Benzoate*. Concise International Chemical Assessment Document No. 26 [Online]. Available: <http://www.whqlibdoc.who.int/publications/2000/924153026X.pdf>. [cited Oct. 17, 2005]

Zhou, X.L., and K. Mopper. 1990. Determination of Photochemically Produced Hydroxyl Radicals in Seawater and Fresh Water. *Marine Chemistry*, 30(1–3):71–88.

ABBREVIATIONS

ACS	American Chemical Society
a.u.	arbitrary units
CAS	Chemical Abstracts Service
DI	deionized
DNA	deoxyribonucleic acid
DOC	dissolved organic carbon
HPLC	high-performance liquid chromatography
IUVA	International Ultraviolet Association
LP	low pressure
MCL	maximum contaminant level
MP	medium pressure
MSDS	material safety data sheet
ND	not determined
NEN	National Standards Institute (the Netherlands)
NOM	natural organic matter
NPOC	nonpurgeable organic carbon
NYC	New York City
REF	reduction equivalent fluence
SA	surface area
TOC	total organic carbon
U.S.	United States
USEPA	U.S. Environmental Protection Agency
USP	U.S. Pharmacopeia
UV	ultraviolet
UVC	ultraviolet C
UVT	ultraviolet transmission
WHO	World Health Organization

IWA Publishing's authorised EU representative for General Product Safety Regulations is Diane D'Arras, 15 rue Duret, 75116 Paris, France, e-mail: safety@iwap.co.uk.

Printed and bound by CPI Group (UK) Ltd, Croydon, CR0 4YY
17/09/2026
02218811-0006